Lecture Notes in Networks and Systems

Volume 45

Series Editor

Janusz Kacprzyk, Systems Research Institute, Polish Academy of Sciences, Warsaw, Poland

Advisory Editors

Fernando Gomide, Department of Computer Engineering and Automation—DCA, School of Electrical and Computer Engineering—FEEC, University of Campinas—UNICAMP, São Paulo, Brazil
Okyay Kaynak, Department of Electrical and Electronic Engineering, Bogazici University, Istanbul, Turkey
Derong Liu, Department of Electrical and Computer Engineering, University of Illinois at Chicago, Chicago, USA; Institute of Automation, Chinese Academy of Sciences, Beijing, China
Witold Pedrycz, Department of Electrical and Computer Engineering, University of Alberta, Alberta, Canada; Systems Research Institute, Polish Academy of Sciences, Warsaw, Poland
Marios M. Polycarpou, KIOS Research Center for Intelligent Systems and Networks, Department of Electrical and Computer Engineering, University of Cyprus, Nicosia, Cyprus
Imre J. Rudas, Óbuda University, Budapest, Hungary
Jun Wang, Department of Computer Science, City University of Hong Kong, Kowloon, Hong Kong

The series "Lecture Notes in Networks and Systems" publishes the latest developments in Networks and Systems–quickly, informally and with high quality. Original research reported in proceedings and post-proceedings represents the core of LNNS.

Volumes published in LNNS embrace all aspects and subfields of, as well as new challenges in, Networks and Systems.

The series contains proceedings and edited volumes in systems and networks, spanning the areas of Cyber-Physical Systems, Autonomous Systems, Sensor Networks, Control Systems, Energy Systems, Automotive Systems, Biological Systems, Vehicular Networking and Connected Vehicles, Aerospace Systems, Automation, Manufacturing, Smart Grids, Nonlinear Systems, Power Systems, Robotics, Social Systems, Economic Systems and other. Of particular value to both the contributors and the readership are the short publication timeframe and the world-wide distribution and exposure which enable both a wide and rapid dissemination of research output.

The series covers the theory, applications, and perspectives on the state of the art and future developments relevant to systems and networks, decision making, control, complex processes and related areas, as embedded in the fields of interdisciplinary and applied sciences, engineering, computer science, physics, economics, social, and life sciences, as well as the paradigms and methodologies behind them.

**** Indexing: The books of this series are submitted to ISI Proceedings, SCOPUS, Google Scholar and Springerlink ****

More information about this series at http://www.springer.com/series/15179

Ludwik Czaja

Cause-Effect Structures

An Algebra of Nets with Examples of Applications

Springer

Ludwik Czaja
Vistula University
Warsaw, Poland

Institute of Informatics
University of Warsaw
Warsaw, Poland

ISSN 2367-3370 ISSN 2367-3389 (electronic)
Lecture Notes in Networks and Systems
ISBN 978-3-030-20460-0 ISBN 978-3-030-20461-7 (eBook)
https://doi.org/10.1007/978-3-030-20461-7

This Springer imprint is published by the registered company Springer Nature Switzerland AG
The registered company address is: Gewerbestrasse 11, 6330 Cham, Switzerland

To Nina and Mateusz

Preface

Cause-effect (c-e) structures are objects of a formal system devised for modeling, testing, and verifying properties of tasks where parallel execution of actions is the most characteristic feature. This is an algebraic system called a *quasi-semiring*, "quasi" because it differs from the standard semiring by somewhat restricted distributivity law, that is $a \cdot (b+c) = a \cdot b + a \cdot c$ fulfilled provided that $b \neq \boldsymbol{\theta}$ if and only if $c \neq \boldsymbol{\theta}$, where $\boldsymbol{\theta}$ is a neutral element for both operations: "multiplication" (denoting simultaneity) and "addition" (denoting nondeterministic choice). Due to the conditional distributivity, this algebra neither reduces to trivial, that is, a single neutral element, nor makes both operations coincide, but describes objects behaviorally equivalent to Petri nets, although differing from them in many respects. The justification of this is in Chaps. 2 and 9. In the graphic presentation, the c-e structures, like Petri nets, are also nets, that is, a kind of graphs, but with one type of nodes, thus not bipartite, as in case of Petri nets.

The main motivation to develop this algebra was to integrate structuring and transformation rules it provides, with graphical and animated presentation of modeled real-world systems—by a special computerized tool. The algebra is a formal basis for combining c-e structures of easy to understand behavior into large system models, some behavioral properties of which can be derived—under well-determined conditions—from behavior of their smaller parts. Such feature, known in the area of programming languages as *compositionality*,[1] is a counterpart of *extensionality* in formal logic. Also, absence of transitions and adjoined to them edges provides more monitor (screen) space for graphic presentation than in case of Petri nets. The appearance on screen is scalable and in many cases uses the so-called "bus layout" (Chaps. 1 and 2), which allows to display some c-e structures of arbitrary size and to avoid excessive entanglement of edges. Apart from the standard issues investigated in the Petri net area, like reachability, liveness,

[1]In some programming languages, breaking the compositionality principle occurs, e.g., in case of the "side effects" of procedures. More general, in [5], the opposite of compositionality is called the "emergent effects". It is known that in accurate mathematical modeling of "real life" systems, the full compositionality, desirable for formal analysis of a model, is often difficult to obtain.

boundedness, etc. (Chap. 8), such problems like lattice properties (Chap. 6) of c-e structures, their decomposition (Chap. 7) or generation of process languages with analysis and synthesis procedures (Chaps. 10–12), are considered.

Cause-effect structures described in the book are an extension of the elementary c-e structures ([2–4]), by the following features: weighted edges, multi-valued nodes having capacities (counterpart of place/transition Petri nets), inhibitors, and several models of time (Chaps. 3 and 4). The extensions are accomplished by modifying the notion of state and semantics, but leaving unchanged structure of the quasi-semiring expressions. Therefore, unlike in case of Petri nets, all these features are included in various definitions of state and semantics, not in the structural (algebraic) definition of c-e structures. The features are illustrated by respective examples. An information system called rough c-e structures is exemplified in Chap. 5.

Chapter 13 contains a number of examples of c-e structures modeling various tasks taken from the "real world". Often their graphic shape resembles appearance of their real-world prototype. In such cases, in the graphic design of a model, the mentioned "bus layout" and symmetries encountered in the modeled objects are applied. So, respective c-e structures are drawn as interconnections of similar (isomorphic) substructures. Adequate examples exhibiting this feature are also in Chaps. 1 through 5.

All the aforesaid examples have been run and tested by means of a computer system comprising graphical editor and simulator of c-e structures [1]. Simulation shows motion of tokens during its run, but sometimes delivers also data about performance of a task being modeled. For instance, experiments made by simulation confirm expectations, how the likelihood of total blockade of a nervous system decreases with increasing its size, i.e., number of neurons, or provide data about dependence between a traffic smoothness of a crossroad and street lights change frequency, given a traffic intensity. A sample of a screen lookout of the editor and simulator is in Chap. 1.

A number of people contributed to the development of elementary c-e structures in the late 1980s and in the next decade, when I have had numerous discussions on the issue, resulting in publications listed at the end of successive chapters of this book and in Related Literature not Referenced in the Chapters. They were: L. Holenderski, A. Szalas, J. Deminet, K. Grygiel, U. Abraham, D. Wikarski, H. W. Pohl, M. Raczunas, A. Maggiolo-Schettini, Nguyen Duc Ha, G. Ciobanu, A. P. Ustimenko, M. Kudlek, who published their contributions, but also others, mainly authors of M.Sc. or Ph.D. dissertations. Also, the students attending my course on selected problems in concurrency at the Institute of Informatics, University of Warsaw, have often been helpful in refining the idea and details. The research on this concept was greatly stimulated by my visits to the Informatik Berichte Humboldt Universitat zu Berlin, Computer Science Department of Ben-Gurion University in Beer-Sheva, Israel, Dipartimento di Informatica

Universita di Pisa, Institute of Informatics Russian Academy of Sciences, Siberian Branch in Novosibirsk and Fachbereich Informatik Universitat Hamburg. The extensions of elementary c-e structures presented in this book being its main subject based on a new concept of state and semantics, arose and have been elaborated in several recent months.

Warsaw, Poland Ludwik Czaja

References

1. Chmielewski RE (2003) Symulacja struktur przyczynowo-skutkowych z wykorzystaniem platformy .NET (Simulation of cause-effect structures using .NET platform), MSc thesis, Institute of Informatics, Warsaw University
2. Czaja L (1988) Cause-effect structures, North-Holland, Information Processing Letters, vol 26, pp 313–319
3. Czaja L (1993) A Calculus of Nets, Kibernetika I Sistemnyj Analiz No.2 March–April 1993 (Russian edition). English edition: Springer, Cybernetics and System Analysis, vol 29, No.2, pp 185–193
4. Czaja L (2002) Elementary cause-effect structures, Warsaw University
5. Hedges J (2016) How to make game theory compositional. In: Proceedings of logic for social behaviour workshop, Zurich

Acknowledgements

Many people contributed to elaboration of the primary version of cause-effect structures, called “elementary” and dating back to the late 1980s. Some of the contributors are mentioned in the Preface. The present version, based on axioms of an algebraic system close to the semiring and encompassing several extensions of the elementary cause-effect structures, has been elaborated in recent months. I am indebted to Krzysztof Czaja for reading the text and calling attention to some of its deficiencies. My special thanks go to Professor Andrzej Skowron for encouragement to the book preparation, and especially for converting its rough appearance—the outcome of my editing—into the style appropriate for publication in Springer-Verlag. I also wish to express my deep thanks to Professor Janusz Kacprzyk, the editor of the “Lecture Notes in Networks and Systems” series, for making possible of its publication.

Ludwik Czaja

Contents

About the Author

Ludwik Czaja obtained his M.Sc., Ph.D., and habilitation degrees from the University of Warsaw, where he was a Professor of Informatics at the Faculty of Mathematics, Informatics and Mechanics. He spent some years in other universities, like Carnegie Mellon, Oxford, Ben-Gurion, Humboldt, as visiting professor or a research fellow. Now he is a Full Professor of the Vistula University and Professor Emeritus of the University of Warsaw. His work encompasses formal and programming languages, compilers, theory of computation, and parallel and distributed processing.

Chapter 1
Introductory Notes

1.1 Motivation

The term "cause-effect structure", is a metaphor taken from phraseology of research branches concerned with concurrency, where phrases like "causality", "effect", "event", "process", "observation", "nondeterminism", "simultaneity", etc. have been adopted. Such terms, partly borrowed from physics by Carl Adam Petri [5], for his net theory, help in the intuitive capture of phenomena investigated in these branches. A similar message lays behind the phenomena dealt with in this book: the combined phrase "cause-effect structure" (c-e structure), emphasises causal relationships between them. This metaphoric phrase refers to the fact that some phenomena take place in effect of their causes, not of their timing interrelationships. The purpose to develop formal background of constructs introduced and illustrated by examples, was to provide means for combining c-e structures, that is models of real-world dynamic objects, into complex large systems, whose behavioural properties might sometimes be inferred from easy to grasp behaviour of their small parts. Such feature is called a *compositionality* (here conditional)—a counterpart of the *extensionality* in formal logic. The means to this end are offered by an algebraic calculus, the so-called *quasi-semiring* of c-e structures. The main motivation of such approach is to combine structuring mechanism and transformation rules it provides, with appeal of simple pictorial and animated presentation of modeled systems. In this presentation, c-e structures are graphs with only one type of nodes, each node endowed with two expressions ("formal polynomials") that determine a way of the nodes interconnection. In some cases, a shape of c-e structures, resembles appearance of the real-world systems being modeled. Examples are road traffic (Figs. 1.1, 13.8, 13.9, 13.10, 13.11, 13.12, 13.13), neural system (Figs. 2.11, 2.12, 2.13, 2.14) or real music text (score) presented in Fig. 4.11, where notes have been used as nodes in respective c-e structure (performance of the score is simulated as activity of the c-e structure). Often c-e structures exhibit symmetry in their graphical representation. It happens when tasks consist of similar parts, differing only by names of nodes, thus respective c-e structures are drawn as combinations of similar (isomorphic) compo-

L. Czaja, *Cause-Effect Structures*, Lecture Notes in Networks and Systems 45, https://doi.org/10.1007/978-3-030-20461-7_1

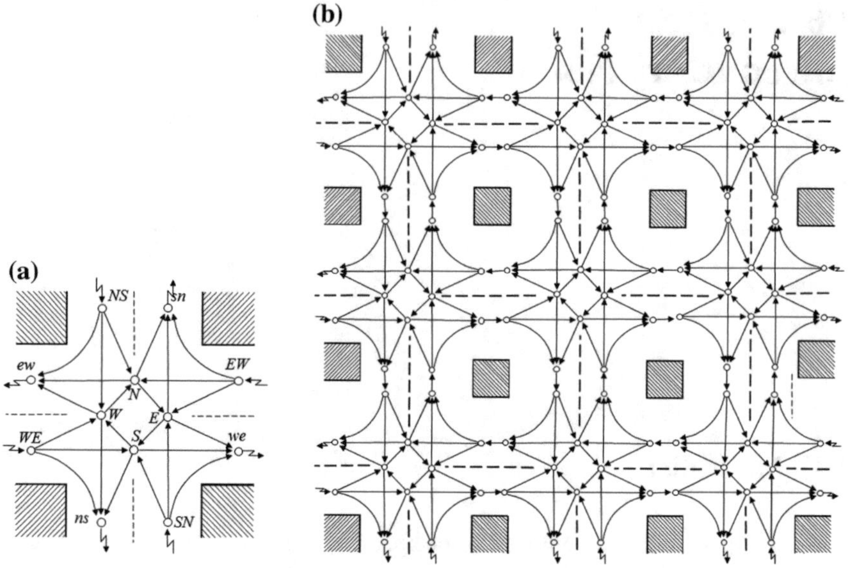

Fig. 1.1 **a** crossroad (NS—entry from the north, ns—exit to the south, etc.; notice the symmetry of the picture: it is composed of four parts of identical shape); resolving collisions and preventing unlawful move of vehicles is enforced in Figs. 13.9, 13.10, 13.11, 13.12, 13.13 (Chap. 13); **b** street grid—a city, arbitrarily extendible

nents. Moreover, for some c-e structures a presentation called a "bus layout", (Fig. 2.4) resembling drawing conventions in computer hardware design, is introduced to avoid excessive entanglement of edges. Examples are an information system of the so-called "rough c-e structures", that specifies some activities during editorial elaboration of documents by a publishing company (Figs. 5.1, 5.2, 5.3) and a system of elevators' motion coordination (Figs. 13.15, 13.16). The bus layout is used when the graphic image of a c-e structure is arbitrarily extendible in the evident way and is a regular combination of similar parts. Such features are shown in Fig. 1.1, where (a) is a schema of a crossroad and (b)—a street grid—multiple copies of (a) that make a regular chart of streets in a city. The complete specification of a crossroad is in Chap. 13.

1.2 Contents of Nodes and Variety of Modeled Task

A node is treated as a location where data reside, operations are executed and decisions taken when performing a task modeled by a c-e structure. Thus, this is not such intention as in case of Petri nets, where places are thought of as conditions or sites of resources, and transitions as events' makers—units transforming state. In a node of the c-e structure a decision is made to which group of successors, called an "effect region", messages will be sent simultaneously and from which group of predecessors, a "cause region"—will be received simultaneously. So, there is a symmetry between dispatch and reception. Various kind of data, like mobile objects (e.g. vehicles, lifts), documents, stimuli, signals, program control points, etc. residing in locations treated as nodes of c-e structures, are represented as tokens - abstractions of the data. Thus, various tasks might be modeled. Examples are: road traffic or motion of other objects with rules of its control, like bridge (Figs. 2.10 and 3.2), crossroad (Figs. 13.8, 13.9, 13.10, 13.11, 13.12, 13.13), lifts (Figs. 13.15, 13.16), boatman (Figs. 13.20, 13.21), neuronal systems functioning (Figs. 2.12), performance of music score (Fig. 4.11), systems with incomplete information (modeled by the so-called here "rough c-e structures", Figs. 5.1, 5.2, 5.3), rules of storing/retrieving data with mutual exclusion (Fig. 3.6), well known tasks illustrating synchronisation (Figs. 13.6, 13.7), rules of communication (Figs. 13.17, 13.18) or enforcement of simultaneous actions (Fig. 13.19). All these example c-e structures have been drawn, run and tested by means of a computer system comprising graphical editor and simulator, devised at the Institute of Informatics, Warsaw University [1]. The simulator shows the tokens' motion during performance of c-e structures, but also has been used for getting some information, like a correlation between a traffic capacity of a crossroad and street lights change frequency, given a traffic intensity.

1.3 Origins

Cause-effect structures considered in this book, evolved from the elementary ones [2] and contain their several extensions. Appearance of the graphic representation is the same in both cases: a c-e structure is a directed graph in which predecessors and successors of each node are grouped into regions of simultaneous reception or dispatch of signals-messages. The difference lays in semantics: various concepts of state and its change, "firing rules", for various extensions of elementary c-e structures.

The definition of elementary c-e structures in their early stage differed from the worked out later [3] but both describing the same objects. Originating in [2] it was recursive, thus a fix-point-oriented, and suggested by the following intended behaviour of the c-e structures. Any node—a working entity like neuron, processor, or whatever, is active (holds a token), when it is <u>ready</u> to send message to its effect region and passive (no token) when <u>ready</u> to receive a message from its cause region. But the readiness does not mean that the node <u>can</u> do this. It depends on whether all the nodes from a certain of its effect (cause) region <u>can</u> receive (send) message from (to) this node. A node <u>can</u> become active if it is passive and every its predecessor from at least one cause region <u>can</u> become passive. And symmetrically: a node <u>can</u> become passive if it is active and every its successor from at least one effect region <u>can</u> become active. This recursive phrase, expressing a symmetry between dispatch and reception, led to equations whose solution defined the semantics of the c-e structure. The solution determines when a node x could change its status, active to passive or conversely, and yields a family of the so-called *impact regions* of x,

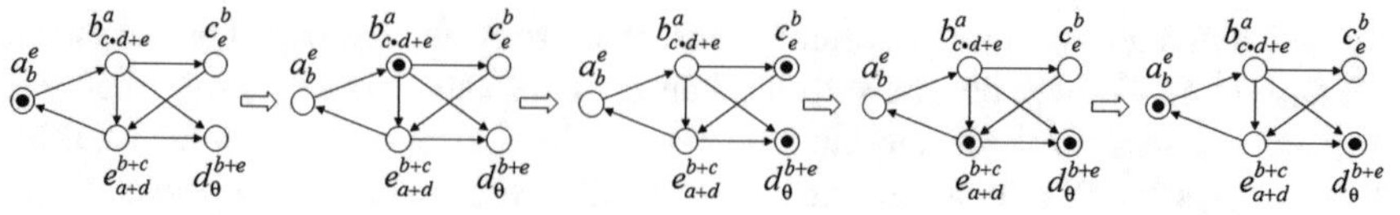

Fig. 1.2 Example of a c-e structure activity

i.e. sets of nodes whose status change should occur along with status change of x. Only one such impact region, called later a *firing component*, was chosen "to fire". The equations were functional in that each node was associated with two boolean functions of state, one for dispatch, one for reception, returning **true**, if the node could change its status in a given state and **false** otherwise. The fix-point approach evolved to the axiomatic one [3, 4], leading to the so-called *quasi-semiring* of c-e structures. Figure 1.2 shows successive transformations of state in a c-e structure, where formal polynomials on the nodes, determine simultaneous (•) or selective (+) reception (upper polynomial) and dispatch (lower polynomial) of tokens.

1.4 Features of the Quasi-semiring of c-e Structures

Apart from the graphical representation, sometimes resembling shape of a modeled system, the c-e structures have two more representations: as a set of nodes endowed with upper and lower polynomials, and as terms (expressions) of the quasi-semiring. For instance, the c-e structure in Fig. 1.2 is a set $\{a^e_b, b^a_{c\bullet d+e}, c^b_e, d^{b+e}_\theta, e^{b+c}_{a+d}\}$ or a term $(a \to b) + (b \to c) \bullet (b \to d) + (b \to e) + (c \to e) + (e \to d) + (e \to a)$. Notice that arrows in the graphical representation are redundant: the interconnection of nodes is determined by the polynomials or structure of respective expression.

The operations $+$ and $\bullet$ on c-e structures are defined as extensions of operations (denoted identically) on formal polynomials, as is illustrated by example in Fig. 1.3.

It is clear how the operations on c-e structures should be defined to imitate composing Petri nets by means of respective operations shown exemplarily in Fig. 1.3c, d, i.e. to get a congruence between calculi of c-e structures and Petri nets. Addition $U + V$: if a node x occurs in U and in V then merge these occurrences into one node x and add up upper polynomials of both occurrences of x and do the same for lower polynomials. Multiplication $U \bullet V$: similarly—replace addition with multiplication of polynomials. This is formalized in Definition 2.3. Operations on formal polynomials are defined axiomatically, yielding the algebraic calculus—a quasi-semiring (Definition 2.1). What deviates this calculus from the ordinary semiring is conditional distribution of multiplication over addition (axiom (+•)) and the same element neutral θ for both operations (axioms (+), (•)). Unlike in case of some (weakened) versions of rings or semirings, the neutral element, being common for both operations does not reduce the whole quasi-semiring to this element, because of conditionality of the distribution law. Another postulate is idempotence of multiplication restricted to single

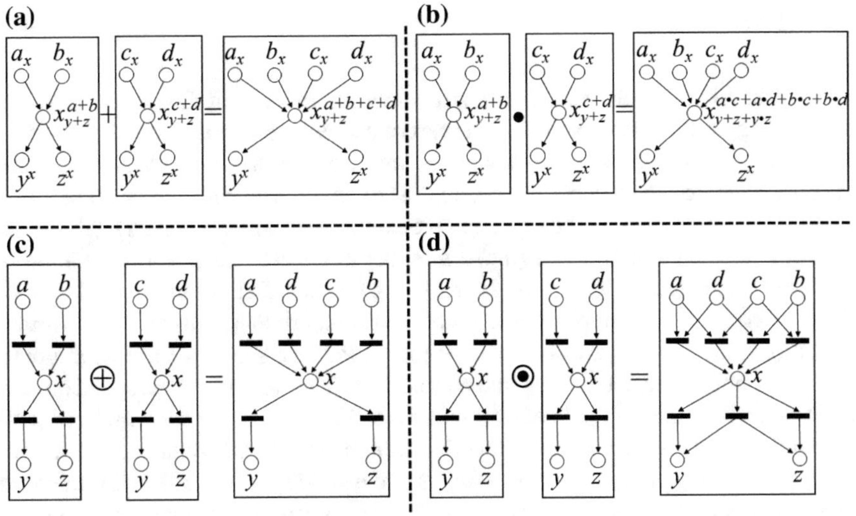

Fig. 1.3 **a** Summation and **b** multiplication of c-e structures; **c**, **d**—similar operations on Petri nets, behaviourally equivalent to c-e structures in (**a**) and (**b**). Operators + and ⊕ make *choice* (of behaviours) between their operands, merging identical nodes occurring in them. Operators • and ⊙ make *synchronisation* (of behaviours) of their operands, merging identical nodes occurring in them

(names of) nodes (axiom (••)). These postulates have been adopted in order to make c-e structures behaviourally equivalent to their Petri net counterparts, as exemplifies Fig. 1.3. The theorem stating this equivalence is in Chap. 9. As mentioned above the c-e structures can be combined in this calculus into complex models of dynamic systems, whose behavioural properties might sometimes be inferred from properties of their substructures where *conditional* compositionality permits (Proposition 8.1, Chap. 8). But this can be decided by checking relevant premisses. Among a number of structural and semantic properties of c-e structures, apart from those being rather direct consequence of the quasi-semiring axioms, there are several somewhat unexpected and not readily verifiable. As examples, are properties related to decomposition of c-e structures (U, distinct from a single arrow, is decomposable if it is a sum or product of structures different from U and θ): there exist infinitely many non-decomposable (*prime*) c-e structures, but every one, after conversion to the so-called two-level one with 3, 4 or 5 nodes, is decomposable (Theorem 7.3, Chap. 7). Also, some lattice properties of infinite c-e structures (Chap. 6), or related to behaviour of Petri nets: the strong and weak equivalence (Chap. 9), required some careful analysis, so that to avoid misleading conclusions drawn up "at the first glance".

1.5 Extensions

Some extensions of elementary c-e structures are defined in Chaps. 3 and 4 by introducing multivalued nodes (counterpart of place/transition Petri nets), inhibitors and time. It has been done by modifying the notion of state and semantics, but leaving syntax (structure) unchanged. Hence, all the structural properties of elementary and extended c-e structures are the same. For the multivalued nodes, the state is a function from nodes to natural numbers (like a marking in p/t Petri nets), for the inhibitors its range is supplemented with ω, symbolising infinity (Definition 3.1 in Chap. 3). Because arrows in the graphic representation are superfluous, thus absent in formal definitions, the counterpart of arrow weights in Petri nets are coefficients in upper and lower polynomials of nodes in c-e structures. The weight ω corresponds to the inhibitors. Examples of application are in Figs. 3.2 and 3.6. For c-e structures with *minimal* time, the state s is also a function from nodes to natural numbers, but with informal meaning: $s(x) = 0$ if no token is at the node x and $s(x) > 1$ is remaining time (a number of ticks in a clock assigned to x) during which the token ***must*** remain at x; $s(x) = 1$ means that time of ***compulsory*** residence of a token at x, expired (Definition 4.1.2 in Chap. 4). For the c-e structures with *maximal* time, replace word "***must***" with "***can***" and "***compulsory***" with "***permissible***" (Definition 4.2.2 in Chap. 4). Thus, $s(x)$ is not meant as number of tokens, but time of token's stay at x. C-e structures with time can be simulated by c-e structures without time constraints, which is illustrated by several examples in Chap. 4. An example of application of sound duration control in music is in Fig. 4.11. Semantics, the "firing rule", for all these categories of c-e structures is accomplished by defining a predicate "***enabled***" for firing component (a minimal substructure without "+" in polynomials) and the next-state relation.

1.6 Processes

Evolution of the c-e structures' activity is represented by their unfoldings in the form of c-e structures (also infinite) with no operation "+" occuring in the polynomials annotating nodes, thus with resolved nondeterminism. The evolutions called processes, with their juxtaposition, i.e. a sequential composition, a kind of concatenation, make monoids of processes. Every process is a concatenation of firing components' occurrences of a c-e structure. But in order to differentiate occurrences of the same firing component in a process, its nodes appear as pairs $\langle$*name, occurrence number*$\rangle$ called events. Events without predecessors bear 0 as occurrence number. Due to the special definition of concatenation, a process represents indeed a particular activity (run) of a c-e structure, recorded as set of events ordered by a precedence relation. Example of a process representing c-e structure's activity shown in Fig. 1.2 is in Fig. 1.4.

Processes and their properties are subject of Chaps. 10, 11, 12.

$$<a,0> \rightarrow <b,1> \begin{matrix} \nearrow <c,1> \rightarrow <e,1> \rightarrow <a,1> \\ \searrow <d,1> \end{matrix}$$

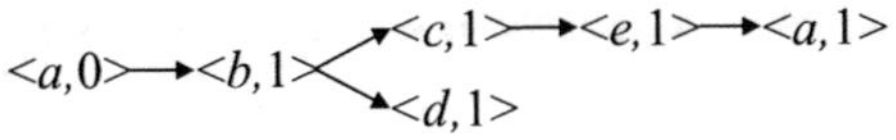

Fig. 1.4   Process—run of the c-e structure from Fig. 1.2

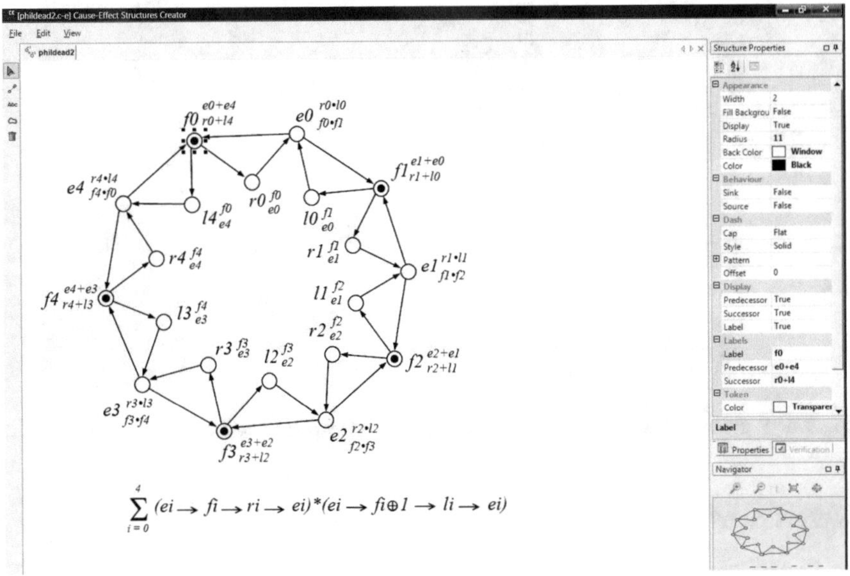

Fig. 1.5   Window of the screen editor with a drawing of example c-e structure

1.7   Screen Editor and Simulator

This tool [1] contains two integrated modules: an editor and simulator. It has been pro-grammed in C# supplied with the .Net platform. The graphical screen editor allows the user to draw graphs of c-e structures on the monitor. There are various functions the editor provides, like pictures' scaling, colours of objects, cut and copy/paste, delete, print, insert and remove tokens, converting image to Powerpoint and to other graphic editors, design of background, and other functions. A sample of the editor's window with image of a c-e structure, is in Fig. 1.5. The simulator takes the c-e struc-ture prepared by the editor on the monitor screen and allows to start its animated activity. There are two modes of its activity: continuous, i.e. automatic and the "sin-gle shot", i.e. by means of consecutive mouse clicks, causing the user's control of passing from state to state. Here too are various functions the simulator provides, like stop of the run, inserting/removing tokens, restore of the initial state, making some nodes disabled or enabled, setting of animation speed and others. Resolving of nondeterministic choice (execution of the "+" operation) is accomplished by a random numbers generator. A sample of the simulator's window during simulation of the c-e structure's run, is shown in Fig. 1.6.

$$\sum_{i=0}^{4} (ei \rightarrow fi \rightarrow ri \rightarrow ei)*(ei \rightarrow fi \oplus 1 \rightarrow li \rightarrow ei)$$

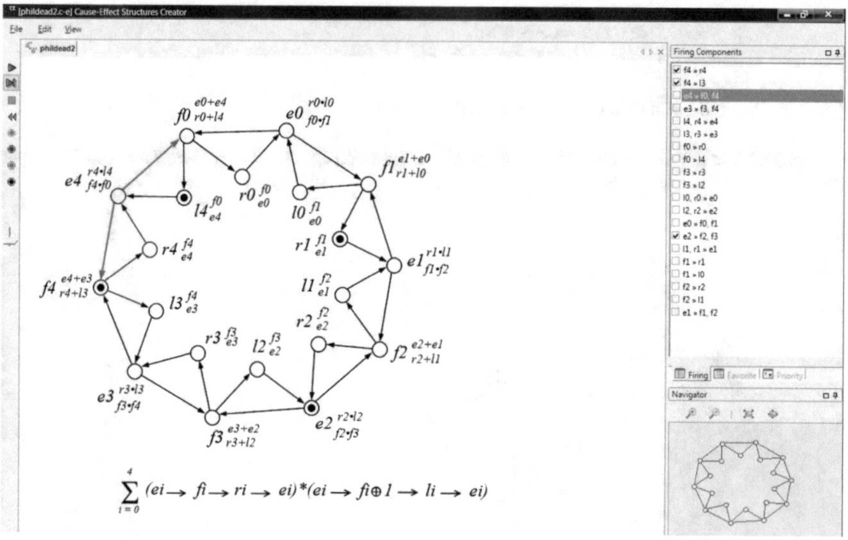

Fig. 1.6   Window of the simulator, a stop-frame during execution of the c-e structure from Fig. 1.5

References

1. Chmielewski RE (2003) Symulacja struktur przyczynowo-skutkowych z wykorzystaniem platformy .NET (Simulation of cause-effect structures using .NET platform), M.Sc. Thesis, Institute
   of Informatics, Warsaw University
2. Czaja L (1988) Cause-effect structures, North-Holland. Inf Process Lett 26:313–319
3. Czaja L (1993) A Calculus of Nets, KIBERNETIKA I SISTEMNYJ ANALIZ No. 2 March-
   April 1993 (Russian edition). English edition: Springer, Cybernetics and System Analysis, vol.
   29, No. 2, pp. 185–193
4. Czaja L (2002) Elementary cause-effect structures. Warsaw University
5. Petri CA (1996) Communication with Automata, Final report, Vol. 1, Supplement 1, RADC
   TR-65-337-vol1-suppl 1, Applied Research, Princeton, NJ. Contract AF 30(602)-3324

Chapter 2
Basics of Elementary Cause-Effect Structures

The elementary cause-effect structures, nets introduced in this chapter, are behaviourally equivalent to condition-event Petri nets [4]. Their constructive features are, however, different. The two main differences are: (1) Each elementary c-e structure has only one type of nodes, the counterparts of Petri net places, their set is infinite, because isolated nodes (without predecessors and successors), though do not affect the structure's behaviour, belong to it. (2) The active objects, unlike Petri net transitions, are not primary notions given explicitly. They are called *firing components* and are determined by two kinds of connections between nodes: one corresponding to non-deterministic choice (represented by operator "+"), the other to simultaneity (represented by operator "•"); the operators combine nodes, making expressions, the so-called formal polynomials and their calculus called a quasi-semiring. The two features allow for direct extension of the operators onto elementary cause-effect structures, making their set also the quasi-semiring.

Definition 2.1 (*set* $\boldsymbol{F}[\mathbb{X}]$, *quasi-semiring of formal polynomials*) Let $\mathbb{X}$ be a non-empty enumerable set. Their elements, called *nodes*, are counterparts of places in Petri nets. Let $\theta \notin \mathbb{X}$ be a symbol called *neutral*. It will play a role of neutral element for operations on terms, called *formal polynomials over* $\mathbb{X}$. The names of nodes, symbol θ, operators +, •, called addition and multiplication, and parentheses are symbols out of which formal polynomials are formed as follows. Each node's name and symbol θ is a formal polynomial; if K and L are formal polynomials, then $(K + L)$ and $(K \bullet L)$ are too; no other formal polynomials exist. Their set is denoted by $\boldsymbol{F}[\mathbb{X}]$. Assume stronger binding of • than +, which allows for dropping some parentheses. Addition and multiplication of formal polynomials is defined as follows: $K \oplus L = (K + L)$, $K \odot L = (K \bullet L)$. To simplify notation, let us use + and • instead of $\oplus$ and $\odot$. It is required that the system $\langle \boldsymbol{F}[\mathbb{X}], +, \bullet, \theta \rangle$ obeys the following equality axioms for all $K, L, M \in \boldsymbol{F}[\mathbb{X}]$, $x \in \mathbb{X}$:

(+)	$\theta + K = K + \theta = K$	(•)	$\theta \bullet K = K \bullet \theta = K$
(++)	$K + K = K$	(••)	$x \bullet x = x$

L. Czaja, *Cause-Effect Structures*, Lecture Notes in Networks and Systems 45, https://doi.org/10.1007/978-3-030-20461-7_2

(+++) $K + L = L + K$ (•••) $K \bullet L = L \bullet K$
(++++) $K + (L + M) = (K + L) + M$ (••••) $K \bullet (L \bullet M) = (K \bullet L) \bullet M$
(+•) If $L \neq \theta \Leftrightarrow M \neq \theta$ then $K \bullet (L + M) = K \bullet L + K \bullet M$

Algebraic system which obeys these axioms will be referred to as a *quasi-semiring of formal polynomials*.[1] □

Remark • The converse implication to that in axiom (+•) is not satisfied. Indeed, if

$$K \bullet (L + M) = K \bullet L + K \bullet M$$

for each K, L, M then $x \bullet (\theta + (x + y)) = x \bullet \theta + x \bullet (x + y)$, thus the equivalence $L \neq \theta \Leftrightarrow M \neq \theta$ is false for $L = \theta$, $M = x + y \neq \theta$.

• Axioms (+) and (•) state that θ is the neutral for both operators. If idempotence of multiplication held for all terms, not only for the single symbols as stated in axiom (••), and if distributivity of • over + held unconditionally, then the two operations would become identical. Indeed, in this case

$$K + L = (K + L) \bullet (K + L) = K + L + K \bullet L,$$

for $K, L \in \boldsymbol{F}[\mathbb{X}]$, but $L + K \bullet L = \theta \bullet L + K \bullet L = (\theta + K) \bullet L = K \bullet L$ and similarly $K + K \bullet L = K \bullet L$. Therefore $K + L = K \bullet L$. This would make this algebraic system inadequate for building cause-effect structures as descriptive tool, alternative and equivalent to Petri nets. In fact, such algebraic system becomes a commutative and idempotent monoid.

• Axiom (++) is equivalent to $\theta + \theta = \theta$. Indeed, (++) $\Rightarrow \theta + \theta = \theta$ follows from (+), and $\theta + \theta = \theta \Rightarrow$ (++) holds since $K = K \bullet \theta = K \bullet (\theta + \theta) = K \bullet \theta + K \bullet \theta = K + K$.

• The axiomatic system $\langle \boldsymbol{F}[\mathbb{X}], +, \bullet, \theta \rangle$ has a model defined as follows: θ is mapped onto {}—the empty family of subsets of $\mathbb{X}$; every $x \in \mathbb{X}$ is mapped onto the family $\{\{x\}\}$ containing only the subset $\{x\}$ of $\mathbb{X}$; if K and L are mapped onto families G and H of subsets of $\mathbb{X}$, then $L + K$ is mapped onto family $G \cup H$ and $K \bullet L$ is mapped onto family:

$$M = \begin{cases} \{g \cup h : \; g \in G \wedge h \in H\} & if \;\; G \neq \{\} \wedge H \neq \{\} \\ G & if \;\; H = \{\} \\ H & if \;\; G = \{\} \end{cases}$$

In this way, each formal polynomial over $\mathbb{X}$ is a notation for a certain finite family of finite subsets of $\mathbb{X}$ and conversely: each such family can be denoted by a certain formal polynomial. For instance, $b \bullet (a + c) + d$ denotes family $\{\{a, b\}, \{b, c\}, \{d\}\}$ of subsets of $\mathbb{X} = \{a, b, c, d, \ldots\}$ (see Fig. 2.3). We use "{}" to denote the empty family

[1] In the first papers on cause-effect structures, the term "near-semiring" has been used. But in the meantime some authors used it in another meaning, so, the term quasi-semiring for this axiomatic system here is adopted.

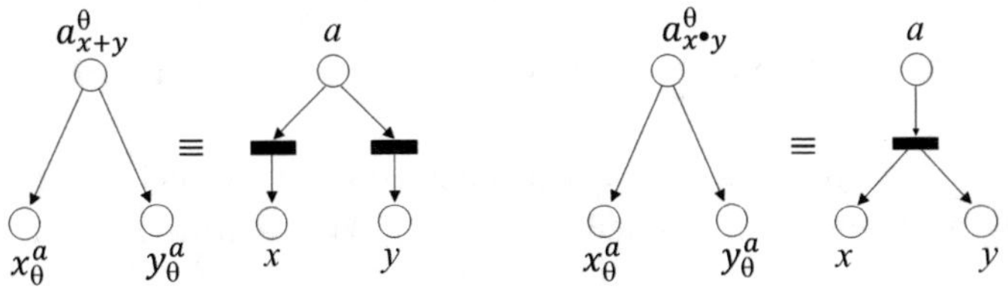

Fig. 2.1 C-e structures and behaviourally equivalent Petri nets

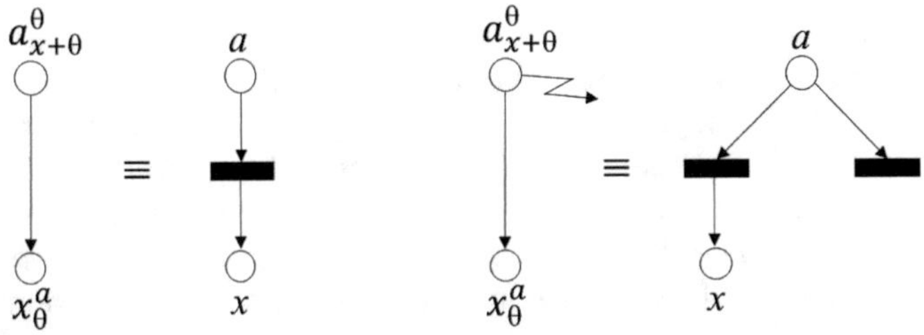

Fig. 2.2 Adding neutral θ–interpretation in terms of Petri nets

of subsets of $\mathbb{X}$, whereas "{{}}" to denote family containing only the empty subset of $\mathbb{X}$. Existence of such model, ensures consistency of the above axiomatic system and is a basis for interpretation of c-e structures as dynamic objects, alternative, but behaviourally equivalent to Petri nets.

• Intended meaning of addition and multiplication is exclusive choice and simultaneity respectively. Thus, for instance, the subscripted node a_{x+y} symbolises the choice between sending token from a to x or to y, whereas $a_{x \bullet y}$ symbolises sending token from a to x and to y simultaneously. So, c-e structures and respective Petri nets in Fig. 2.1 should act identically.

Anticipating formal definitions of operations on c-e structures, the c-e structure at the left side in Fig. 2.1 is the sum and at the right side - product of arrows $\{a^{\theta}_{x}, x^{a}_{\theta}\}$ and $\{a^{\theta}_{y}, y^{a}_{\theta}\}$. For $y = \theta$, two possible meanings (in terms of Petri nets) of adding the neutral symbol are in Fig. 2.2.

There are two reasons of adopting meaning expressed by the left equivalence in Fig. 2.2. (1) Nets with "dangling" transitions have no direct equivalents amongst c-e structures; nonetheless their behaviour equivalent to that of Petri nets, may be simulated and symbolised pictorially by a zigzag arrow, as at the right side of Fig. 2.2. (2) The subscript (superscript) θ of a node symbolises lack of the node's successors (predecessors); thus, exclusive choice between $\{a^{\theta}_{x}, x^{a}_{\theta}\}$ and the empty c-e structure (defined later as $\{a^{\theta}_{\theta}\}$, being the isolated node a), should result in sending token from a to x, but not exclusively removing it from a. General discussions on interpretation of various kinds of choice is in [1–3, 5].

• A useful remark for calculations with formal polynomials, concerns their transforming into the canonical form, i.e. a sum of products of node symbols: each polynomial $K \in \boldsymbol{F}[\mathbb{X}]$, $K \neq \theta$, may be transformed by equations (+)–(+•) in Definition

2.1, into the canonical form $\sum_{i=1}^{n}\bigwedge_{j=1}^{k_i} x_{ij}$ where x_{ij} is a node symbol occurring in K ($\sum$ and $\bigwedge$ are operators of repetitive summation by "+" and multiplication by "•"). Easy proof—by structural induction: let L and M be in canonical form, if $K = L + M$ or $K = L \bullet M$ then use appropriate equations (+)–(+•) and conclude that K is in the canonical form.

Definition 2.2 (*cause-effect structure, carrier, set* ***CE***) A cause-effect structure (c-e structure) over $\mathbb{X}$ is a pair $U = (C, E)$ of total functions:
$C\colon\ \mathbb{X} \to \boldsymbol{F}[\mathbb{X}]$ (*cause function*; nodes occuring in $C(x)$ are *causes* of x)
$E\colon\ \mathbb{X} \to \boldsymbol{F}[\mathbb{X}]$ (*effect function*; nodes occuring in $E(x)$ are *effects* of x)
such that x occurs in the formal polynomial $C(y)$ iff y occurs in $E(x)$. *Carrier* of U is the set $car(U) = \{x \in \mathbb{X} : C(x) \neq \theta \vee E(x) \neq \theta\}$. U is of *finite carrier* iff $|car(U)| < \infty$ ($|\ldots|$ denotes cardinality). The set of all c-e structures over $\mathbb{X}$ is denoted by $\boldsymbol{CE}[\mathbb{X}]$. Since $\mathbb{X}$ is fixed, write just ***CE***—wherever this makes no confusion. □

Since functions C and E are total, each c-e structure comprises all nodes from $\mathbb{X}$, also the isolated ones—those from outside of its carrier. Presenting c-e structures graphically, only their carriers are pictured.

A representation of a c-e structure $U = (C, E)$ as a set of annotated nodes is $\{x^{C(x)}_{E(x)} : x \in car(U)\}$. U is also presented as a directed graph with $car(U)$ as set of vertices labelled with objects of the form $x^{C(x)}_{E(x)}$ ($x \in car(U)$); there is an edge from x to y iff y occurs in the polynomial $E(x)$. Note that in this representation, edges (arrows), although useful for the graphical look-out, are redundant: interconnection of vertices may be inferred from polynomials $C(x)$, $E(x)$. In some examples we shall omit $C(x)$ or $E(x)$ if they are θ. Since functions C, E are total, any c-e structure comprises all the nodes from $\mathbb{X}$, also the isolated ones (with $C(x) = E(x) = \theta$), invisible in the graphical representation. The isolated nodes, in fact, make the distributivity law (+•) conditional. Fig. 2.3a, b depict two graphical presentations of the same c-e structure with carrier $\{a, b, c, d, e, f, g, h\}$; in (a) the encircled nodes comprise groups making products in formal polynomials in (b),

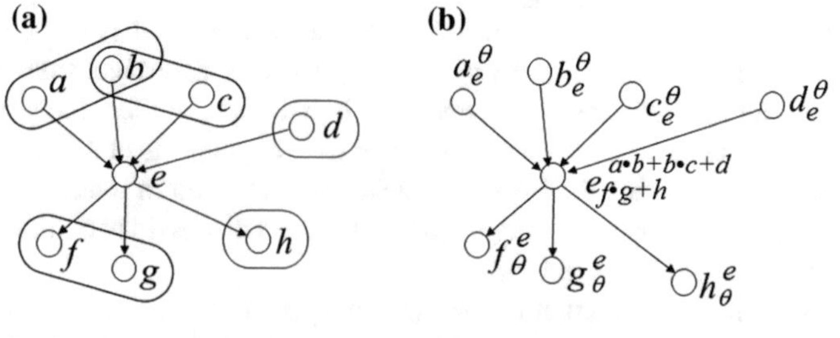

Fig. 2.3 **a** Predecessors and successors of the node e, grouped into families: $\{\{a, b\}, \{b, c\}, \{d\}\}$ and $\{\{f, g\}, \{h\}\}$. **b** Notation by means of polynomials

where the sums of the products create families of the groups. This c-e structure is the set $\{a_e^\theta, b_e^\theta, c_e^\theta, d_e^\theta, e_{f\cdot g+h}^{a\cdot b+b\cdot c+d}, f_\theta^e, g_\theta^e, h_\theta^e\}$.

Definition 2.3 (*addition and multiplication, monomial c-e structure*) For c-e structures $U = (C_U, E_U)$, $V = (C_V, E_V)$ let us define:

(i) $U + V = (C_{U+V}, E_{U+V}) = (C_U + C_V, E_U + E_V)$,
where $(C_U + C_V)(x) = C_U(x) + C_V(x)$ and similarly for E

(ii) $U \bullet V = (C_{U\bullet V}, E_{U\bullet V}) = (C_U \bullet C_V, E_U \bullet E_V)$,
where $(C_U \bullet C_V)(x) = C_U(x) \bullet C_V(x)$ and similarly for E

(The same symbols "+" and "•" are used for operations on c-e structures and formal polynomials)

U is a *monomial* c-e structure iff each polynomial $C_U(x)$ and $E_U(x)$ is a monomial, i.e. does not comprise non-reducible (relative to equations in Definition 2.1) operation "+". □

Evidently $U + V \in \boldsymbol{CE}$ and $U \bullet V \in \boldsymbol{CE}$ that is, in the resulting structures, x occurs in $C_{U+V}(y)$ iff y occurs in $E_{U+V}(x)$ and the same for $U \bullet V$. Thus, addition and multiplication of c-e structures yield correct c-e structures.

2.1 Representation of c-e Structures by Expressions Built Up Out of Arrows

Apart from the representation of c-e structures as a set

$$\{x_{E(x)}^{C(x)} : x \in car(U)\}$$

their linear notation is used as the so-called "*arrow-expressions*": $\{x_y^\theta, y_\theta^x\}$ is an arrow, denoted as $x \to y$ and, consequently, $\{x_y^\theta, y_\theta^x\} \bullet \{y_z^\theta, z_\theta^y\} \bullet \{z_u^\theta, u_\theta^z\} \ldots =$ $\{x_y^\theta, y_z^x, z_u^y, u_{\ldots}^z \ldots\}$ is a chain, denoted as $x \to y \to z \to u \ldots$. Bidirectional arrow $x \leftrightarrow y$ denotes $x \to y \to x$ (equivalent to $y \to x \to y$), that is, the close cycle $\{x_y^y, y_x^x\}$. Chains and arrows in particular, may be combined into "arrow expressions" representing some c-e structures. For instance c-e structure $\{a_{x+y}^\theta, b_{x\bullet y}^\theta, x_\theta^{a\bullet b}, y_\theta^{a\bullet b}\}$ may be written as $(a \to x + a \to y) \bullet (b \to x) \bullet (b \to y)$.

2.2 Drawing Patterns

To make some layouts of graphic presentation of examples clearer, let us admit some drawing patterns used in some examples. The layouts in Fig. 2.4 allow for avoiding excessive entanglement of arrows: they may be called "bus layouts", since they resemble some drawing conventions in computer hardware design.

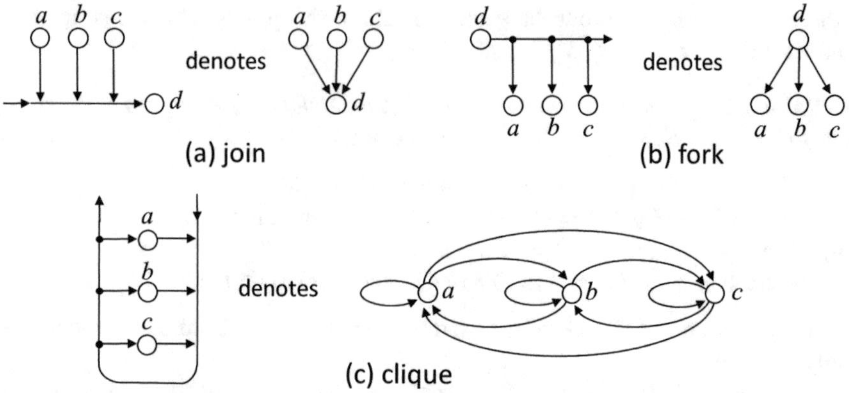

Fig. 2.4 Patterns of "bus layouts"

The set ***CE*** with addition, multiplication and a distinguished element denoted also by θ and understood as the empty c-e structure (θ, θ), where θ is a constant function $\theta(x) = \theta$ for all $x \in \mathbb{X}$, makes an algebraic system similar to that in Definition 2.1.

Proposition 2.1 (quasi-semiring of c-e structures) *The system* $\langle \boldsymbol{CE}[\mathbb{X}], +, \bullet, \theta \rangle$ *obeys the following equations for all* $U, V, W \in \boldsymbol{CE}[\mathbb{X}]$, $x, y \in \mathbb{X}$:

(+) $\theta + U = U + \theta = U$ — ($\bullet$) $\theta \bullet U = U \bullet \theta = U$

(++) $U + U = U$ — ($\bullet\bullet$) $(x \to y) \bullet (x \to y) = x \to y$

(+++) $U + V = U + V$ — ($\bullet\bullet\bullet$) $U \bullet V = V \bullet U$

(++++) $U + (V + W) = (U + V) + W$ — ($\bullet\bullet\bullet\bullet$) $U \bullet (V \bullet W) = (U \bullet V) \bullet W$

(+$\bullet$) *If* $C_V(x) \neq \theta \Leftrightarrow C_W(x) \neq \theta$ *and* $E_V(x) \neq \theta \Leftrightarrow E_W(x) \neq \theta$ *then* $U \bullet (V + W) = U \bullet V + U \bullet W$ □

This follows directly from definition of c-e structures and definitions of adding and multiplying c-e structures. The operations on c-e structures make possible to combine small c-e structures into large ones. Due to associativity stated in (++++) and ($\bullet\bullet\bullet\bullet$) the parentheses may be omitted, thus $\sum$ and $\bigwedge$ are used to denote repetitive summation and multiplication of c-e structures. An example of indispensability of the premise for distributivity of multiplication over addition stated in (+$\bullet$) is illustrated in Fig. 2.5, because $U \bullet (V + W) \neq U \bullet V + U \bullet W$ in this case.

Definition 2.4 (*partial order* $\leq$; *substructure, set* ***SUB***$[V]$) For $U, V \in \boldsymbol{CE}$ let $U \leq V \Leftrightarrow V = U + V$; obviously, $\leq$ is a partial order in ***CE***. If $U \leq V$ then U is a *substructure* of V; $\boldsymbol{SUB}[V] = \{U : U \leq V\}$ is the set of all substructures of V. For $A \subseteq \boldsymbol{CE}$:
$V \in A$ is *minimal* (w.r.t. $\leq$) in A iff $\forall W \in A$: $(W \leq V \Rightarrow W = V)$ □

The crucial notion for behaviour of c-e structures is firing component, a counterpart of transition in Petri nets, i.e. a state transformer. It is, however, not a primitive

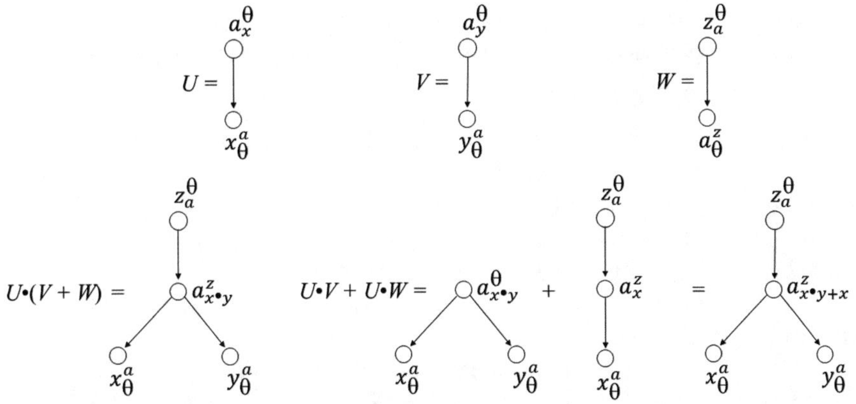

Fig. 2.5 A token residing at the node a would go to x and y simultaneously in the c-e structure $U \bullet (V + W)$, whereas to either x and y simultaneously, or only to x in the c-e structure $U \bullet V + U \bullet W$

notion but derived from the definition of c-e structures, and is introduced regardless of any particular c-e structure:

Definition 2.5 (*firing component, set* $\boldsymbol{FC}$, *pre-set and post-set*) A minimal in $\boldsymbol{CE} \backslash \{\theta\}$ c-e structure $Q = (C_Q, E_Q)$ is a *firing component* iff Q is a monomial c-e structure and $C_Q(x) = \theta \Leftrightarrow E_Q(x) \neq \theta$ for any $x \in car(Q)$. The set of all firing components is denoted by $\boldsymbol{FC}$, thus the set of all firing components of $U \in \boldsymbol{CE}$ is $\boldsymbol{FC}[U] = \boldsymbol{SUB}[U] \cap \boldsymbol{FC}$. Following the standard Petri net notation, let for $Q \in \boldsymbol{FC}$ and $G \subseteq \boldsymbol{FC}$:

$$
\begin{aligned}
&{}^{\bullet}Q = \{x \in car(Q) : C_Q(x) = \theta\} && (\textit{pre-set} \text{ of } Q)\\
&Q^{\bullet} = \{x \in car(Q) : E_Q(x) = \theta\} && (\textit{post-set} \text{ of } Q)\\
&{}^{\bullet}Q^{\bullet} = {}^{\bullet}Q \cup Q^{\bullet} &&\\
&{}^{\bullet}G = \bigcup_{Q \in G} {}^{\bullet}Q && (\textit{pre-set}\text{of } G)\\
&G^{\bullet} = \bigcup_{Q \in G} Q^{\bullet} && (\textit{post-set} \text{ of } G)\\
&{}^{\bullet}G^{\bullet} = {}^{\bullet}G \cup G^{\bullet} &&
\end{aligned}
$$

□

Notice that the firing component is a connected graph, due to the required minimality. Elements of the pre-set are its *causes* and elements of the post-set are its *effects*.

Some immediate, easy to verify consequences of above definitions are in:

Proposition 2.2 *For any c-e structures* $U_1, V_1, U_2, V_2, U, V, W$:
(*a*) $U_1 \leq V_1 \wedge U_2 \leq V_2 \Rightarrow U_1 + U_2 \leq V_1 + V_2$ (*monotonicity of* $+$)
(*b*) *If* $(U_1 + V_1) \bullet (U_2 + V_2) = U_1 \bullet U_2 + V_1 \bullet V_2 + U_1 \bullet V_2 + V_1 \bullet U_2$ *then* $U_1 \leq V_1 \wedge U_2 \leq V_2 \Rightarrow U_1 \bullet U_2 \leq V_1 \bullet V_2$ (*conditional monotonicity of* $\bullet$)
(*c*) $U \bullet (V + W) \leq U \bullet V + U \bullet W$
(*d*) *If* $U \bullet (V + W) = U \bullet V + U \bullet W$ *then* $V \leq W \Rightarrow U \bullet V \leq U \bullet W$

(e) $(U \bullet V)^n = U^n \bullet V^n$ *(Define:* $U^0 = \theta,\ U^{n+1} = U \bullet U^n$*), for* $n = 0, 1, 2, \ldots$.
(f) Let $C_U(x) \neq \theta \Leftrightarrow C_V(x) \neq \theta$ *and* $E_U(x) \neq \theta \Leftrightarrow E_V(x) \neq \theta$ *for any* x.
Then $(U + V)^n = \sum_{i=0}^{n} U^{n-i} \bullet V^i$
(g) If $car(U) \cap car(V) = \emptyset$ *and* $V \leq W$ *then* $V \leq U \bullet W$
(h) $U \leq V \Leftrightarrow \boldsymbol{SUB}[U] \subseteq \boldsymbol{SUB}[V]$
(i) $\boldsymbol{SUB}[U] \cup \boldsymbol{SUB}[V] \subseteq \boldsymbol{SUB}[U + V]$
(j) Inclusion "$\subseteq$*" in (i) cannot, in general, be replaced with equality "*$=$*"*
(k) $\boldsymbol{SUB}[U] \cup \boldsymbol{SUB}[V] = \boldsymbol{SUB}[U + V] \Leftrightarrow (U \leq V \vee V \leq U)$
(l) $U \leq V \Rightarrow \boldsymbol{FC}[U] \subseteq \boldsymbol{FC}[V]$ *but converse implication not always holds*
(m) $\boldsymbol{FC}[U] \cup \boldsymbol{FC}[V] \subseteq \boldsymbol{FC}[U + V]$
(n) Inclusion "$\subseteq$*" in (m) cannot, in general, be replaced with equality "*$=$*"*
(o) $\boldsymbol{SUB}[U] \cup \boldsymbol{SUB}[V] = \boldsymbol{SUB}[U + V] \Rightarrow \boldsymbol{FC}[U] \cup \boldsymbol{FC}[V] = \boldsymbol{FC}[U + V]$
but converse implication not always holds
(p) Any firing component $Q \in \boldsymbol{FC}$ *is a directed and connected bipartite graph with k arrows beginning in the set* $^\bullet Q$ *and ending in* $Q^\bullet$*, where k satisfies* $|^\bullet Q| + |Q^\bullet| - 1 \leq k \leq |^\bullet Q| \cdot |Q^\bullet|$ □

Points (m) and (n) in Proposition 2.2 are worth some comments. Point (n) states that new firing components may appear when summing up c-e structures. For instance, let $U = \{a_{x+y}, b_{x\bullet y}, x^{a\bullet b}, y^{a\bullet b}\}$, $V = \{a_{x\bullet y}, x^a, y^a\}$, thus $\boldsymbol{F}\mathrm{C}[U] = \emptyset$, $\boldsymbol{F}\mathrm{C}[V] = \{V\}$, $\boldsymbol{F}\mathrm{C}[U + V] = \{\{a_x, x^a\}, \{a_y, y^a\}, V, \{a_{x\bullet y}, b_{x\bullet y}, x^{a\bullet b}, y^{a\bullet b}\}\}$, thus $\boldsymbol{F}\mathrm{C}[U] \cup \boldsymbol{F}\mathrm{C}[V] \neq \boldsymbol{F}\ \mathrm{C}[U + V]$. The phenomenon of creating new firing components when assembling c-e structures from smaller parts, reflects a general observation: compound systems may sometimes reveal behaviours absent in their parts. However, the systems obtained by summation, retain behaviour of their parts.

Definition 2.6 (*state of elementary c-e structures*) A *state* is a subset of the set of nodes: $s \subseteq \mathbb{X}$. The set of all states: $\mathbb{S} = 2^{\mathbb{X}}$. A node x is *active* at the state s iff $x \in s$ and *passive* otherwise. As in Petri nets phraseology we say "x holds a token" when x is active. Obviously, the state might be defined equivalently as a two-valued function $s\colon \mathbb{X} \to \{0, 1\}$. □

Definition 2.7 (*semantics* [[ ]] *of elementary c-e structures*) For $Q \in \boldsymbol{FC}[U]$, $s, t \in \mathbb{S}$, let $[[Q]] \subseteq \mathbb{S} \times \mathbb{S}$ be a binary relation defined as: $(s, t) \in [[Q]]$ iff $^\bullet Q \subseteq s \wedge Q^\bullet \cap s = \emptyset \wedge t = (s \setminus {^\bullet Q}) \cup Q^\bullet$ (say: Q transforms state s into t). *Semantics* $[[U]]$ of $U \in \boldsymbol{CE}$ is: $[[U]] = \bigcup_{Q \in FC[U]} [[Q]]$. $[[U]]^*$ is its reflexive and transitive closure, that is, $(s, t) \in [[U]]^*$ if and only if $s = t$ or there exists a sequence of states $s_0, s_1, \ldots, s_n$ with $s = s_0$, $t = s_n$ and $(s_j, s_{j+1}) \in [[U]]$ for $j = 0, 1, \ldots, n - 1$. We say that t is *reachable* from s in semantics [[ ]] and the sequence $s_0, s_1, \ldots, s_n$ is called a *computation* in U. □

Notice that $[[U]] = \emptyset$ if and only if $\boldsymbol{FC}\ [U] = \emptyset$. Informally, behaviour of c-e structures in accordance with this semantics may be imagined as a token game: if each node in a firing component's pre-set holds a token and none in its post-set does, then remove tokens from the pre-set and put them in the post-set.

Definition 2.8 (*parallel semantics* $[[\,]]_{par}$ *of elementary c-e structures*) Firing components Q and P are *detached* if and only if $^{\bullet}Q^{\bullet} \cap {}^{\bullet}P^{\bullet} = \emptyset$. For $G \subseteq \boldsymbol{FC}[U]$, $G \neq \emptyset$, containing only pairwise detached (*pwd*) firing components, let $[[G]]_{par} \subseteq \mathbb{S} \times \mathbb{S}$ be a binary relation defined as: $(s, t) \in [[G]]_{par}$ iff ${}^{\bullet}G \subseteq s \wedge G^{\bullet} \cap s = \emptyset \wedge t = (s \backslash {}^{\bullet}G) \cup G^{\bullet}$ (say: G transforms state s into t). *Semantics* $[[U]]_{par}$ of $U \in \boldsymbol{CE}$ is: $[[U]]_{par} = \bigcup_{G \subseteq \boldsymbol{FC}[U] \wedge G \text{ is } pwd} [[G]]_{par}$. Closure $[[U]]^{*}_{par}$ and reachability and computation are defined as in Definition 2.7. □

Remark Any firing component is a two level bipartite connected graph with $n > 0$ causes and $m > 0$ effects. Firing components with the same pre-set and the same post-set have identical semantics, thus behave identically: if ${}^{\bullet}Q = {}^{\bullet}P$ and $Q^{\bullet} = P^{\bullet}$ then $[[Q]] = [[P]]$. Q and P are behaviourally equivalent. For instance, there are 5 equivalent firing components with $n = 2$ and $m = 2$, these are the following $\{a^{\theta}_{x \bullet y}, b^{\theta}_{y}, x^{a}_{\theta}, y^{a \bullet b}_{\theta}\}$, $\{a^{\theta}_{x}, b^{\theta}_{x \bullet y}, x^{a \bullet b}_{\theta}, y^{b}_{\theta}\}$, $\{a^{\theta}_{y}, b^{\theta}_{x \bullet y}, x^{b}_{\theta}, y^{a \bullet b}_{\theta}\}$, $\{a^{\theta}_{x \bullet y}, b^{\theta}_{x}, x^{a \bullet b}_{\theta}, y^{a}_{\theta}\}$, $\{a^{\theta}_{x \bullet y}, b^{\theta}_{x \bullet y}, x^{a \bullet b}_{\theta}, y^{a \bullet b}_{\theta}\}$. The last of them will be called "closely connected" (Definition 8.1).

Some consequences of above definitions are in Proposition 2.3. Most of them are almost obvious, several, a little less evident and used in further chapters, are proved in Chap. 8.

Proposition 2.3 *For any firing components* $Q, P \in \boldsymbol{FC}$, *non-empty sets* $G, H \subseteq \boldsymbol{FC}$ *of pairwise detached firing components, and any c-e structures* $U, V \in \boldsymbol{CE}$:
(a) $[[Q]] \neq \emptyset$, $[[G]]_{par} \neq \emptyset$
(b) $[[Q]] \neq [[P]] \Leftrightarrow [[Q]] \cap [[P]] = \emptyset$, $[[G]]_{par} \neq [[H]]_{par} \Leftrightarrow [[G]]_{par} \cap [[H]]_{par} = \emptyset$
(c) $U \leq V \Rightarrow \boldsymbol{FC}[U] \subseteq \boldsymbol{FC}[V] \Rightarrow [[U]] \subseteq [[V]] \Rightarrow [[U]]^{*} \subseteq [[V]]^{*}$ *but none of the implications "$\Rightarrow$" may be replaced with equivalence "$\Leftrightarrow$"*
(d) $\boldsymbol{FC}[U] \subseteq \boldsymbol{FC}[V] \Rightarrow [[U]]_{par} \subseteq [[V]]_{par} \Rightarrow [[U]]^{*}_{par} \subseteq [[V]]^{*}_{par}$ *but none of the implications "$\Rightarrow$" may be replaced with equivalence "$\Leftrightarrow$"*
(e) $[[U]] \subseteq [[U]]_{par}$ *but the reverse inclusion not always holds (proof in Proposition 8.1)*
(f) $[[U]]^{*} = [[U]]^{*}_{par}$ *(proof in Proposition 8.2)*
(g) $[[U]] \cup [[V]] \subseteq [[U + V]]$, *but the reverse inclusion not always holds (proof in Proposition 8.2)*
(h) $[[U]] \cup [[V]] \subseteq [[U]]^{*} \cup [[V]]^{*} \subseteq ([[U]] \cup [[V]])^{*} \subseteq [[U + V]]^{*}$ *but none of the inclusions "$\subseteq$". may be replaced with equality "$=$"*
(i) $\boldsymbol{FC}[U] \cup \boldsymbol{FC}[V] = \boldsymbol{FC}[U + V] \Rightarrow [[U]] \cup [[V]] = [[U + V]]$ *but not conversely (proof in Proposition 8.1)*
(j) $\boldsymbol{FC}[U] \cup \boldsymbol{FC}[V] = \boldsymbol{FC}[U + V]$ *and* $[[U]]^{*} \cup [[V]]^{*} = [[U + V]]^{*}$ *are unrelated by implication.*
(k) $\boldsymbol{FC}[U] \cup \boldsymbol{FC}[V] = \boldsymbol{FC}[U + V]$ *and* $[[U]]_{par} \cup [[V]]_{par} = [[U + V]]_{par}$ *are unrelated by implication.*
(l) $\boldsymbol{FC}[U] \cup \boldsymbol{FC}[V] = \boldsymbol{FC}[U + V]$ *and* $[[U]]^{*}_{par} \cup [[V]]^{*}_{par} = [[U + V]]^{*}_{par}$ *are unrelated by implication.*

□

Remark Point (g) states that composition of some c-e structures may show behaviour absent in its constituents. For instance, if $U = \{a_{x+y}, b_{x \bullet y}, x^{a \bullet b}, y^{a \bullet b}\}$, $V = \{a_{x \bullet y}, x^{a}, y^{a}\}$, then $[[U]] = \emptyset$, $[[V]] = [[\{a_{x \bullet y}, x^{a}, y^{a}\}]]$, $[[U + V]] = [[\{a_{x \bullet y}, x^{a}, y^{a}\}]] \cup [[\{a_{x}, x^{a}\}]] \cup [[\{a_{y}, y^{a}\}]] \cup [[\{a_{x \bullet y}, b_{x \bullet y}, x^{a \bullet b}, y^{a \bullet b}\}]]$. Thus, $[[U]] \cup [[V]] \neq [[U + V]]$. This means that, the semantics, in general, is not fully compositional, but point (i) gives a sufficient condition for being such, proved in Proposition 8.1.

2.3 Example (Bridge)

A two-lane road—each for vehicles heading in the direction opposite to the other—passes a bridge capable of hosting at most one vehicle at a time, as in Figs. 2.6, 2.7. Let us select five labelled points on each lane:

Point *B* represents the bridge, *E*, *W* are points on entrance to the bridge and *e*, *w*—on exit from the bridge. Primed points are direct neighbours of respective unprimed. Another representation of the lanes is by sets of superscripted and subscripted (labels of) points:

$$EW = \{E'^{\theta}_{E},\ E^{E'}_{B},\ B^{E}_{e},\ e^{B}_{e'},\ e'^{e}_{\theta}\}$$
$$WE = \{W'^{\theta}_{W},\ W^{W'}_{B},\ B^{W}_{w},\ w^{B}_{w'},\ w'^{w}_{\theta}\}$$

Let us combine the two lanes into the road as follows: $ROAD = EW + WE$, where "+" means making union of the two sets, with formal "sum" of superscripts and subscripts of points identically labelled in both summand-sets. So, $ROAD$ is a sum of two c-e structures according to Definition 2.3:

$$ROAD = \{E'^{\theta}_{E},\ E^{E'}_{B},\ e^{B}_{e'},\ e'^{e}_{\theta},\ B^{E+W}_{e+w},\ W'^{\theta}_{W},\ W^{W'}_{B},\ w^{B}_{w'},\ w'^{w}_{\theta}\}$$

Pictorially, *ROAD* is presented as a graph in Fig. 2.8.

Vehicles-tokens travel from point to point along the arrows and each point hosts at most one token at a time. Thus (1) the addition of superscripts or subscripts represents nondeterministic choice between the two addends, so, *B* takes a token either from *E* or from *W* and sends either to *e* or to *w*; (2) some undesirable moves, the U-turns, are not prevented: $E' \to E \to B \to w \to w'$ and $W' \to W \to B \to e \to e'$. To prevent these moves, let us introduce two small structures $R = B \longleftrightarrow r$ and

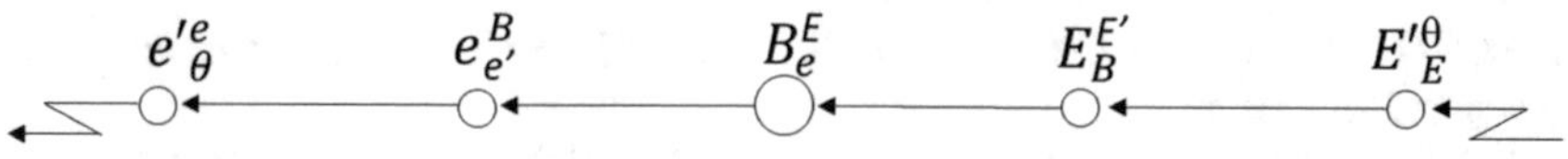

Fig. 2.6 $EW = E' \to E \to B \to e \to e'$ East-West lane

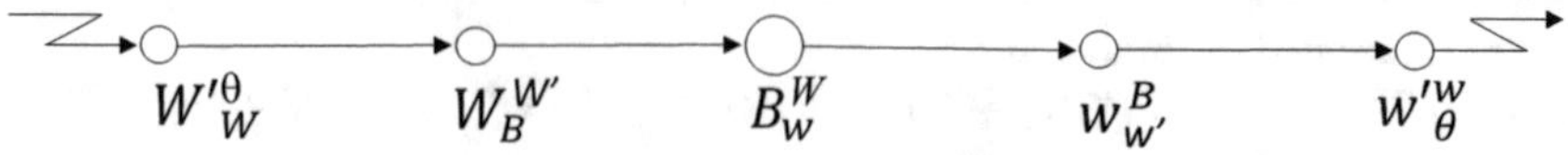

Fig. 2.7 $WE = W' \to W \to B \to w \to w'$ West-East lane

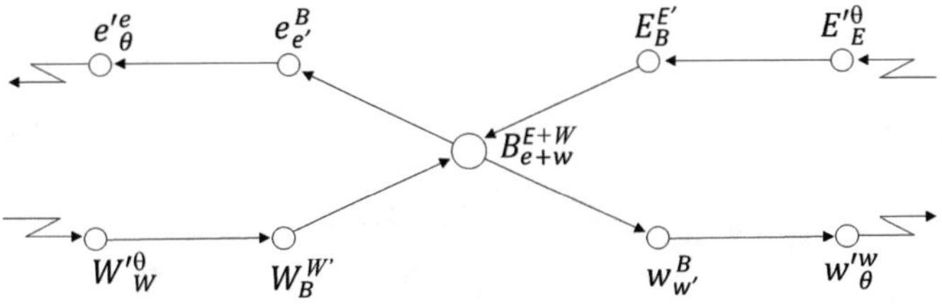

Fig. 2.8 Unlawful moves (U-turn on the Bridge B possible)

Fig. 2.9 Control of direction

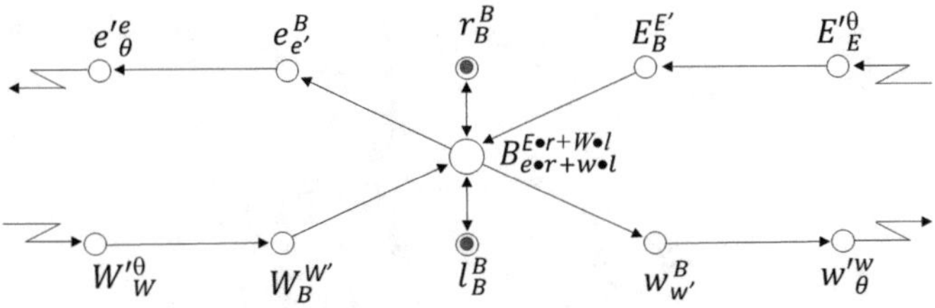

Fig. 2.10 Two way road traffic controlled: no U-turn on the bridge B permitted

$L = B \longleftrightarrow l$, shown in Fig. 2.9, represented also by sets: $R = \{B^{r}_{r},\ r^{B}_{B},\}$ and $L = \{B^{l}_{l},\ l^{B}_{B},\}$ (nodes r and l might be imagined as traffic red lights or semaphores).

Now, let us combine them with the former as follows:

$ROAD1 = EW \bullet R + WE \bullet L$, where "$\bullet$" means making union of the two sets, with formal "product" of superscripts and subscripts of points identically labelled in both multiplicant-sets, presented as a graph in Fig. 2.10.

2.4 Example (Neurons)

A simplified schematic hand drawing of a biological neuron (a neural cell), is depicted in Fig. 2.11 and exemplary c-e structure representing a neuron is in Fig. 2.12.

Some or all of dendrite entry points a, b, c, d are activated, by external signals arriving from axon exit points of adjacent neurons. Junctions of axon exit points with dendrite entry points are synapses. The signals arrive in dendrite entry points and depart the neuron through axon exit points. Let us assume that only with the neuron body B (but not with the synapses), a certain number τ_B called the B's ***threshold*** is associated. We also neglect "direct synapses": junctions of axon exit points with neuron bodies. Another simplifying assumption is that the potential of instantaneous entry (to B) is measured by a number of active dendrite entry points, hence, each

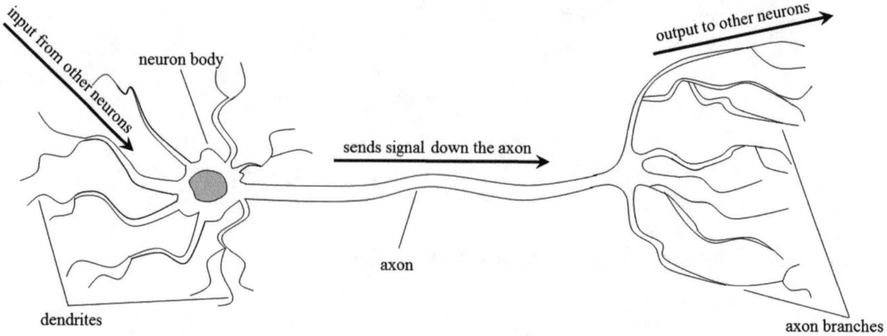

Fig. 2.11 A simplified vision of a neuron

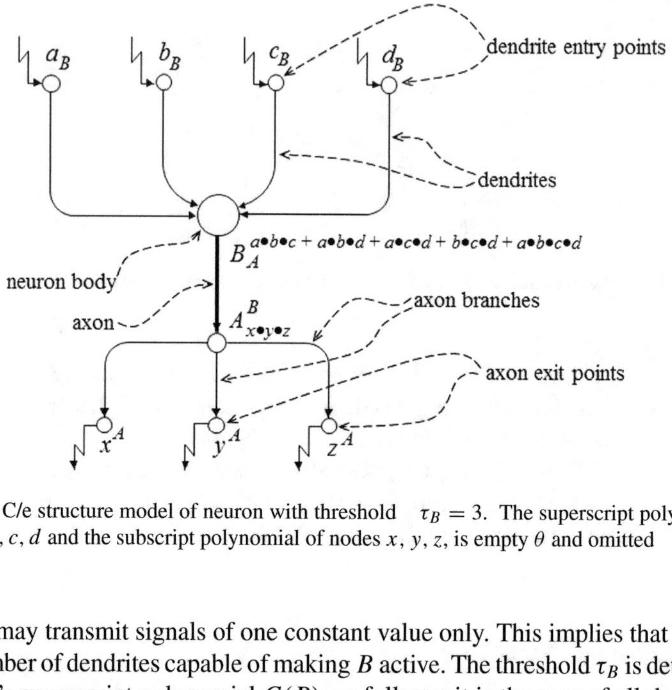

Fig. 2.12 C/e structure model of neuron with threshold $\tau_B = 3$. The superscript polynomial of nodes a, b, c, d and the subscript polynomial of nodes x, y, z, is empty θ and omitted

synapse may transmit signals of one constant value only. This implies that τ_B is the least number of dendrites capable of making B active. The threshold τ_B is determined by the B's superscript polynomial $C(B)$, as follows: it is the sum of all the distinct monomials, each being a product of at least τ_B distinct dendrite entry point names. In Fig. 2.12, $\tau_B = 3$ thus,
$C(B) = a \bullet b \bullet c + a \bullet b \bullet d + a \bullet c \bullet d + b \bullet c \bullet d + a \bullet b \bullet c \bullet d.$
In general, if there are n dendrites entering a neuron body B then polynomial $C(B)$ has

$$\sum_{k=\tau_B}^{n} \binom{n}{k} = \sum_{k=\tau_B}^{n} \frac{n!}{k!(n-k)!}$$ products having respectively,

$\tau_B, \tau_B + 1, .\tau_B + 2, \ldots, n$ factors. The neuron body loses its activation by sending a signal through the axon to its branches that begin in one branching point, in Fig. 2.12 denoted by A and ending with the axon exit points denoted by x, y, z. From A

the signal distributes evenly among all the axon exit points. From the latter points, the signals depart the neuron. Since the neuron has one body only, then we will refer to any neuron by the name of its body.

Specification of the neuron in Fig. 2.12 by means of arrow expression is given by the following equations:

$DENDRITES = (a \to B) \bullet (b \to B) \bullet (c \to B) + (a \to B) \bullet (b \to B) \bullet (d \to B) + (a \to B) \bullet (c \to B) \bullet (d \to B) + (b \to B) \bullet (c \to B) \bullet (d \to B) + (a \to B) \bullet (b \to B) \bullet (c \to B) \bullet (d \to B)$

$AXON = B \to A$

$AXONBRANCHES = (A \to x) \bullet (A \to y) \bullet (A \to z)$

$NEURON = DENDRITES \bullet AXON \bullet AXONBRANCHES$

2.5 Neuron's Body Threshold Control

In Fig. 2.12 a neuron cell with the fixed value of threshold has been shown. A specification, of the same neuron cell with modifiable threshold τ_B, is depicted in Fig. 2.13. The maximal value of the threshold is 4. Nodes **1, 2, 3, 4** (bold), are used to control the threshold value: inserting a token in one of them, yields the threshold value 1 or 2 or 3 or 4, respectively. The superscript polynomial of the node B is

$$\mathbf{1} \bullet (a + b + c + d) + \mathbf{2} \bullet (a \bullet b + a \bullet c + a \bullet d + b \bullet c + b \bullet d + c \bullet d) + \mathbf{3} \bullet (a \bullet b \bullet c + a \bullet b \bullet d + a \bullet c \bullet d + b \bullet c \bullet d) + \mathbf{4} \bullet (a \bullet b \bullet c \bullet d).$$

Figure 2.14 depicts a c-e structure containing three neurons with stabilized thresholds of the values $\tau_{B1} = 3,\ \tau_{B2} = 2,\ \tau_{B3} = 4$.

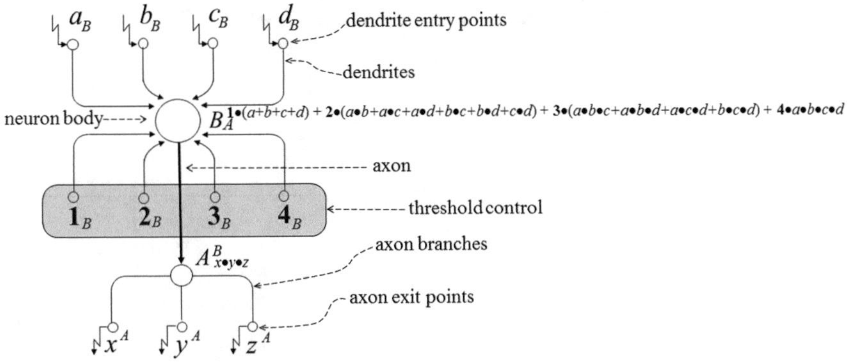

Fig. 2.13 A neuron with modifiable body threshold. Stimuli (tokens) from the environment, are arriving in control nodes on the shadowed fragment. For instance, a token in **2** determines $\tau_B = 2$. The environment modifies the threshold value, which may tend to stabilization of these values in some neurons, leading to a sort of "learning" of the c-e structure

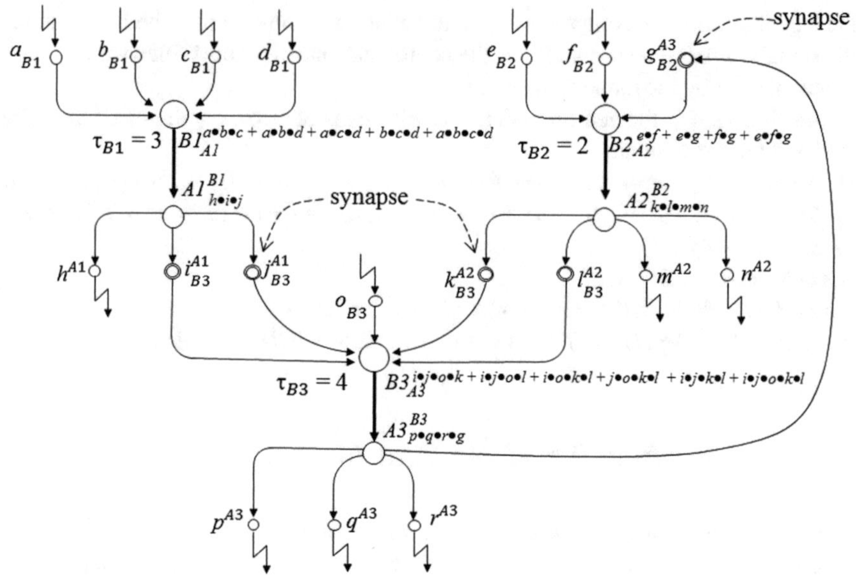

Fig. 2.14 A state of thresholds in the network of three neurons

References

1. Bruns G (1997) Distributed systems analysis with CCS. Prentice-Hall
2. Hoare CAR (1985) Communicating sequential processes. Prentice Hall
3. Milner R (1989) Communication and concurrency. Prentice Hall Inc
4. Reisig W (1985) Petri nets. An introduction, Number 4 in EATCS monographs on theoretical computer science. Springer, Berlin, Heidelberg, New York, Tokyo
5. Roscoe AW (1998) The theory and practice of concurrency. Prentice-Hall

Chapter 3
Extensions of Elementary Cause-Effect Structures: Inhibitors, Multi-valued Nodes with Capacities and Weighted Monomials

Each elementary c-e structure has been defined as a pair of total functions with a fixed (one and the same for all of them), infinite set $\mathbb{X}$ of nodes as the domain and the set $\boldsymbol{F}[\mathbb{X}]$ of formal polynomials over $\mathbb{X}$ as the range. This made possible straightforward definition of their addition and multiplication, i.e. quasi-semiring of c-e structures, without additional and artificial assumptions on their different domains and ranges. The same concerned general definition of firing component as a particular c-e structure, not as a part of a larger c-e structure. Although the structural definition of firing component is the same as in Definition 2.5, its semantics may be different, as states Definition 3.3 that follows. In this section it is assumed that with a given (i.e. already constructed) c-e structure $U \in \boldsymbol{CE}[\mathbb{X}]$ and the set of its firing components $\boldsymbol{FC}[U] = \boldsymbol{SUB}[U] \cap \boldsymbol{FC}$, some additional information is associated. The following extensions of c-e structures endowed with such information will be acquired: c-e structures with multi-valued nodes and their capacities, weights of arrows (so, counterparts of place/transition Petri nets) and with inhibitors. To this end, a notation for multisets is convenient.

3.1 A Notation for Multisets

Let $\mathbb{N}$ be the set of all natural numbers including 0 and $\mathbb{N}_\omega = \mathbb{N} \cup \{\omega\}$, where the value ω symbolises infinity, that is $\omega > n$ for each $n \in \mathbb{N}$. A *multiset* over a set X is a (total) function $f\colon X \to \mathbb{N}_\omega$. If the set $\{x\colon f(x) \neq 0\}$ is finite then the linear-form notation is adopted for multisets , e.g.

X	a	b	c	d	e
$f(X)$	2	0	3	1	ω

is denoted as $2 \otimes a + 3 \otimes c + d + \omega \otimes e$. A multiset is *zero* $\mathbb{O}$, when $\mathbb{O}(x) = 0$ for all x. Addition, subtraction and multiplication on multisets is defined: $(f + g)(x) = f(x) + g(x)$, $(f - g)(x) = f(x) - g(x)$ for $f(x) \geq g(x)$, $(f \cdot g)(x) = f(x) \cdot g(x)$,

L. Czaja, *Cause-Effect Structures*, Lecture Notes in Networks and Systems 45, https://doi.org/10.1007/978-3-030-20461-7_3

comparison of multisets: $f \leq g$ iff $f(x) \leq g(x)$ for all x. Assume the customary arithmetic of ω: $\omega + n = \omega$, $\omega - n = \omega$, $\omega + \omega = \omega$ and additionally $0 - \omega = 0$ needed in Definition 3.4.

The state of extended c-e structures is redefined:

Definition 3.1 (*state of c-e structure*) A state of extended c-e structure U is a total function $\bar{s} : car(U) \to \mathbb{N}_\omega$, thus a multiset over $car(U)$. The set of all states of U is denoted by $\mathbb{S}$. □

Definition 3.2 (*weights of monomials and capacity of nodes*) Given a c-e structure $U = \langle C, E \rangle$ and its firing component $Q \in \boldsymbol{FC}[U]$, let along with the pre-set $^\bullet Q$ and post-set $Q^\bullet$ of Q, some multisets $\overline{^\bullet Q}$: $^\bullet Q \to \mathbb{N}_\omega \backslash \{0\}$ and $\overline{Q^\bullet}$: $Q^\bullet \to \mathbb{N}_\omega \backslash \{0\}$ be given as additional information. The value $\overline{^\bullet Q}(x)$ is called a *weight* (or *multiplicity*) of monomial $E_Q(x)$ and the value $\overline{Q^\bullet}(x)$—a *weight* (or *multiplicity*) of monomial $C_Q(x)$. Additionally, let $\overline{^\bullet Q}(x) = 0$ for $x \notin {}^\bullet Q$ and $\overline{Q^\bullet}(x) = 0$ for $x \notin Q^\bullet$. Let *cap* be a total injective function $cap : car(U) \to \mathbb{N}_\omega \backslash \{0\}$, assigning a *capacity* to each node in the set $car(U)$. A c-e structure with such enhanced firing components is called a *c-e structure-with-weights of monomials and capacity of nodes*. □

Definition 3.3 (*firing components enabled and with inhibitors*) For a firing component $Q \in \boldsymbol{FC}[U]$ let us define the set

$$inh[Q] = \{x \in {}^\bullet Q : \overline{^\bullet Q}(x) = \omega\},$$

that is the set of nodes in the pre-set of Q, whose effect monomials $E_Q(x)$ are of weight ω (by Definition 2.5 $E_Q(x) \neq \theta \Leftrightarrow C_Q(x) = \theta$)— see Fig. 3.1. The nodes in $inh[Q]$ will play role of *inhibiting nodes* of firing component Q, as follows. For Q and state $\bar{s}$ let us define the formula: $enabled[Q](\bar{s})$ if and only if:

$$\begin{aligned}
&\forall x \in inh[Q] : \bar{s}(x) = 0 \wedge \\
&\forall x \in {}^\bullet Q \backslash inh[Q] : \overline{^\bullet Q}(x) \leq \bar{s}(x) \leq cap(x) \wedge \\
&\forall x \in Q^\bullet : \overline{Q^\bullet}(x) + \bar{s}(x) \leq cap(x)
\end{aligned}$$

In words: Q is enabled at the state $\bar{s}$ iff none of inhibiting nodes $x \in {}^\bullet Q$ contains a token and each remaining node in $^\bullet Q$ does, with no fewer tokens than is the weight of its effect monomial $E_Q(x)$ and no more than capacity of each $x \in {}^\bullet Q$. Moreover, none of $x \in Q^\bullet$ holds more tokens than their number increased by the weight of its cause monomial $C_Q(x)$, exceeds capacity of x. The inhibiting nodes of a firing component will be called its *inhibitors*. □

A weighted effect monomial of a node $a \in {}^\bullet Q$ is $\overline{^\bullet Q}(a) \otimes E_Q(a)$, denoted by $\overline{E}_Q(a)$, where $E_Q(a)$ is not weighted one. Similarly for the weighted cause monomial of a node $x \in Q^\bullet$: $\overline{Q^\bullet}(x) \otimes C_Q(x)$, denoted by $\overline{C}_Q(x)$, where $C_Q(x)$ is not weighted one. Figure 3.1a shows a firing component Q with weighted effect monomials $\overline{E}_Q(a) = 5 \otimes x$, $\overline{E}_Q(b) = \omega \otimes (x \bullet y)$, $\overline{E}_Q(c) = 3 \otimes y$ and weighted cause monomials $\overline{C}_Q(x) = 2 \otimes (a \bullet b)$, $\overline{C}_Q(y) = 4 \otimes (b \bullet c)$. The inhibitor of Q

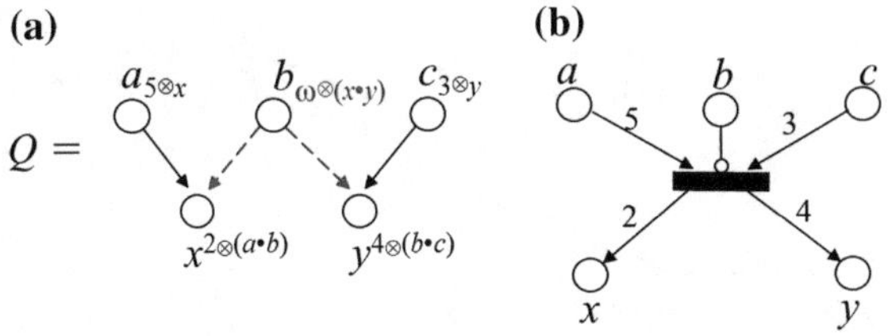

Fig. 3.1 **a** Firing component Q with weights; for instance $2 \otimes (a \bullet b)$, $\omega \otimes (x \bullet y)$, etc. denote multiplicity of the product $a \bullet b$ by the factor $2 = \overline{Q^{\bullet}}(x)$ and the product $x \bullet y$ by the factor $\omega = \overline{{}^{\bullet}Q}(b)$. **b** Behaviourally equivalent Petri net transition

is the node b. The red dashed arrows play inhibiting role. Figure 3.1b shows the behaviourally equivalent single transition in Petri net with weights and inhibitor arrow.

Now, we are in a position to define semantics ("firing rule") for all classes of cause-effect structures:

Definition 3.4 (*semantics* [[ ]] *of extended c-e structures*) For $Q \in \boldsymbol{FC}[U]$, let $[[Q]] \subseteq \mathbb{S} \times \mathbb{S}$ be a binary relation defined as:
$(\bar{s}, \bar{t}) \in [[Q]]$ iff $enabled[Q](\bar{s}) \wedge \bar{t} = (\bar{s} - \overline{{}^{\bullet}Q}) + \overline{Q^{\bullet}} \leq cap$ (say: Q transforms state $\bar{s}$ into $\bar{t}$). *Semantics* $[[U]]$ of $U \in \boldsymbol{CE}$ is $[[U]] = \bigcup_{Q \in \boldsymbol{FC}[U]} [[Q]]$. Closure $[[U]]^*$ and reachability and computation are defined as in Definition 2.7. □

Definition 3.5 (*parallel semantics* $[[\ ]]_{par}$ *of extended c-e structures*) For $G \subseteq \boldsymbol{FC}[U]$, $G \neq \emptyset$, containing only pairwise detached (*pwd*) firing components, let $[[G]]_{par} \subseteq \mathbb{S} \times \mathbb{S}$ be a binary relation defined as: $(\bar{s}, \bar{t}) \in [[G]]_{par}$ iff $\forall Q \in G : (enabled[Q](\bar{s}) \wedge \bar{t} = (\bar{s} - \overline{{}^{\bullet}Q}) + \overline{Q^{\bullet}} \leq cap)$ (say: G transforms state $\bar{s}$ into $\bar{t}$). *Semantics* $[[U]]_{par}$ of $U \in \boldsymbol{CE}$ is: $[[U]]_{par} = \bigcup_{G \subseteq \mathbf{FC}[U] \wedge G \text{ is } pwd} [[G]]_{par}$. Closure $[[U]]^*_{par}$, reachability and computation are defined as in Definition 2.7. □

Remark Semantics of any kind of c-e structure U is defined as the union of relations $[[Q]] \subseteq \mathbb{S} \times \mathbb{S}$ regardless of how this relation is constructed. Therefore semantical consequences, for instance, of interrelationship between sets $\boldsymbol{FC}[U]$, $\boldsymbol{FC}[V]$, $\boldsymbol{FC}[U + V]$, for c-e structures introduced in this chapter, are the same as in Proposition 2.3. Hence the following:

Proposition 3.1 *Proposition 2.3 is valid also for c-e structures with multi-token nodes, with weights in upper and lower polynomials annotating nodes and with inhibitors. That is, for extensions of elementary c-e structures.* □

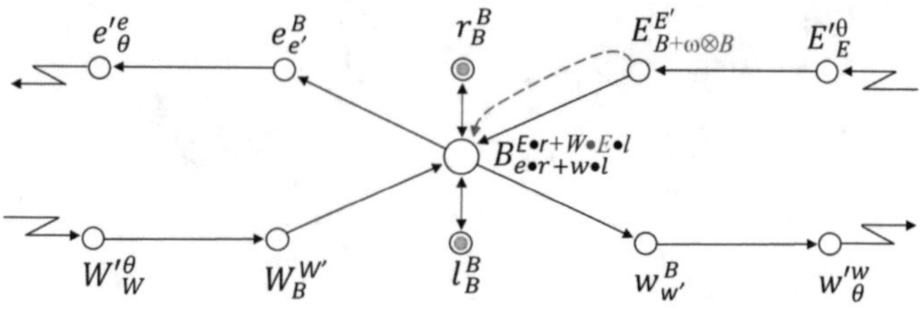

Fig. 3.2 If at E and W are vehicles (tokens) and none at B, then only the one in E will get entry permit at B, since only firing component $\{E^{\theta}_{B}, r^{\theta}_{B}, B^{E \bullet r}_{\theta}\}$ can fire in such state, not this one: $\{W^{\theta}_{B}, E^{\theta}_{\omega \otimes B}, l^{\theta}_{B}, B^{W \bullet E \bullet l}_{\theta}\}$ –due to its inhibiting node E if it contains a token. So, the node E plays two roles: ordinary and inhibiting, depending on a fire component it belongs to. If no vehicles are in E and W then r and l (nodes preventing U-turn) contain tokens

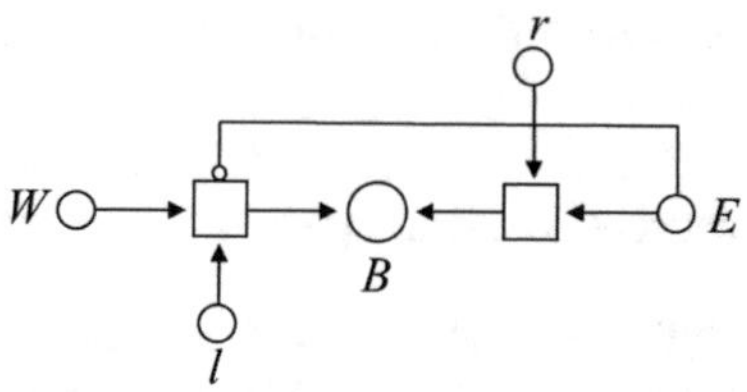

Fig. 3.3 Petri net counterpart of two firing components with place B of c-e structure shown in Fig. 3.2. Inhibitor arc leads from place E to the left transition

3.2 Example (Bridge with Priority to Ride from the East)

In the c-e structure in Fig. 2.9, which presents a ride through the bridge B, the priority ride from the East can be enforced using inhibitor, i.e. node E in the pre-set of firing component $\{W^{\theta}_{B}, E^{\theta}_{\omega \otimes B}, l^{\theta}_{B}, B^{W \bullet E \bullet l}_{\theta}\}$, as shown in Fig. 3.2.

Firing components $\{E^{\theta}_{B}, r^{\theta}_{B}, B^{E \cdot r}_{\theta}\}$ and $\{W^{\theta}_{B}, E^{\theta}_{\omega \otimes B}, l^{\theta}_{B}, B^{W \cdot E \cdot l}_{\theta}\}$ of the c-e structure in Fig. 3.2 have Petri nets (with inhibitor arc) counterparts as two transitions shown in Fig. 3.3.

Evidently:

- **the elementary c-e structures** (counterparts of Condition/Event Petri nets) are obtained if the range of the state is reduced to $\{0, 1\}$, $\overline{{}^{\bullet}Q}(x) = \overline{Q^{\bullet}}(x) = 1$, $cap(x) = 1$ for each firing component Q and each node x; then the formula $enabled[Q](\bar{s})$ reduces to $\forall x \in {}^{\bullet}Q : \bar{s}(x) = 1 \wedge \forall x \in Q^{\bullet} : \bar{s}(x) = 0$
- **the c-e structures with capacity of nodes and weights-of-monomials** (counterparts of P/T Petri nets) are obtained if the formula $enabled[Q](\bar{s})$ reduces to $\forall x \in {}^{\bullet}Q : \overline{{}^{\bullet}Q}(x) \leq \bar{s}(x) \leq cap(x) \wedge \forall x \in Q^{\bullet} : \overline{Q^{\bullet}}(x) + \bar{s}(x) \leq cap(x)$
- **the c-e structures with** $inh[Q] \neq \varnothing$ **and the formula** $enabled[Q](\bar{s})$ **as in Definition** 3.3, are counterparts of P/T Petri nets with inhibiting arrows.

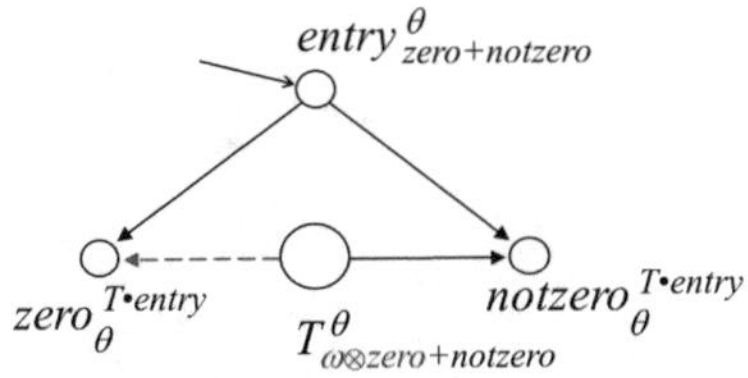

Fig. 3.4 Test zero performed by extended c-e structure with inhibitor node T. Node $entry$ starts testing contents of node T: if it is empty then a token goes to $zero$, otherwise—to $notzero$

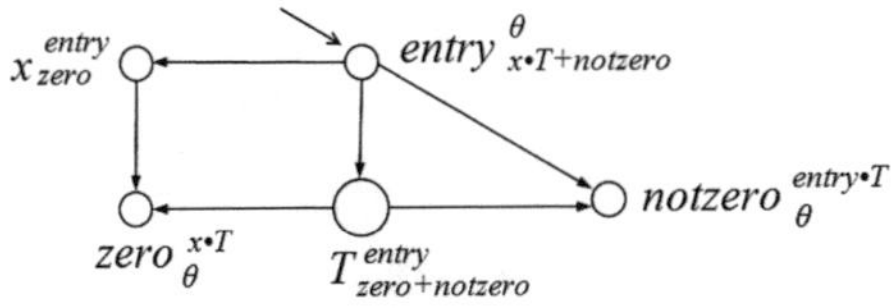

Fig. 3.5 Test zero performed by elementary c-e structure without inhibitors. Node $entry$ starts testing contents of node T. If T is empty, a token goes to $zero$, otherwise—to $notzero$

3.3 Example (Test zero)

This is a c-e structure which detects emptiness (presence or absence of a token) of a given node. "Test zero" is possible for the multi-valued c-e structures with weights, arbitrary capacity of nodes and with inhibitors, but not possible without inhibitors. However the "test zero" is possible for the elementary c-e structures (their nodes hold at most one token and have no inhibitors). This is shown in Figs. 3.4 and 3.5.

It is known that ordinary P/T Petri nets, are incapable of checking if places of unbounded capacity, contain tokens, but become capable if are extended with inhibitor arcs, obtaining then the full Turing power. The same concerns extended c-e structures considered in this chapter. This "test zero" is possible for elementary c-e structures but at the price of additional nodes, as shows Fig. 3.5.

3.4 Example (Readers/Writers Problem)

The widely known example of "Readers/Writers" originates in [1]. A set of n sequential agents (programs) run concurrently under the constraint: writing to a common file by the jth ($j = 1, 2, \ldots, n$) agent prevents all remaining agents from reading and writing, but not from their private (internal) activity. Reading may proceed in parallel. Figure 3.6 shows three agents with the following meaning of nodes: Aj—agent of number $j = 1, 2, 3$ is active (holds a token) if it is neither reading nor writing; Rj—is active if the jth agent is reading; Wj—is active if the jth agent is writing. Wj and Rj play both roles: of the ordinary nodes or of the inhibitors, dependently which firing component they belong to. Verification of correctness (safety) of the c-e structure in Fig. 3.6, relative to its specification (by a formula), is in Chap. 8.

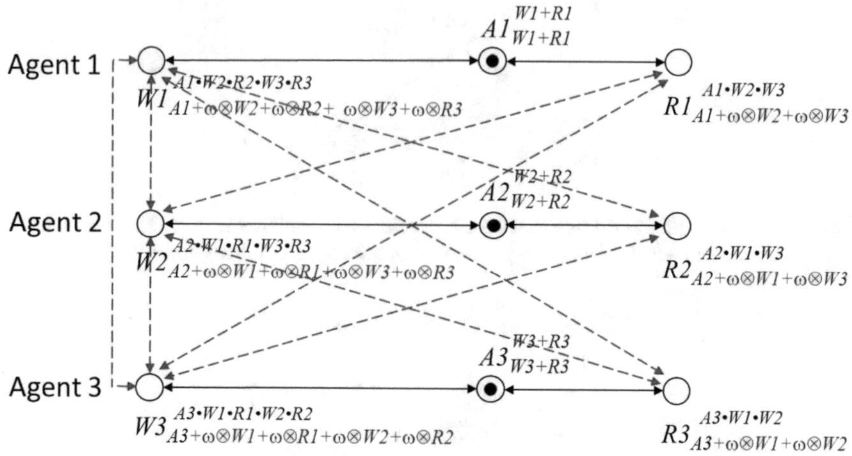

Fig. 3.6 Three agents' Readers/Writers system model as a c-e structure RW with inhibitors; the dashed red arrows denote usage of inhibitors. Initially, the agents are neither reading nor writing (tokens at $A1$, $A2$, $A3$)

Reference

1. Brinch Hansen P (1973) Operating system principles. Prencice-Hall, Inc., Englewood Cliffs, New Jersey, USA

Chapter 4
Another Extension: Time in Elementary c-e Structures

The time models considered here are different from those found in the literature on the Petri nets with time: time usually treated as a period of activity of a transition, here has been separated into minimal and maximal time required for a token to reside in a given node.

4.1 Minimal Time Model

In the minimal time model, with each node a time period of ***mandatory*** stay of a token at this node is associated. This is the shortest time during which the node ***must*** hold the token. On expiry of this period, the token ***can*** leave the node (if other necessary conditions for this "move" are met). Time flow may be related either to:

(a) individual firing components or to

(b) the whole c-e structure.

In the case (a), called a ***local min-time model***, time is measured by separate timers (each with perhaps diverse frequency, thus advancement rate of time) assigned to nodes of a given firing component only, or by a timer shared by these nodes; the timers do not impact any action outside this firing component.

In the case (b), called a ***global min-time model***, time is measured by an external (global) timer referred to by nodes of the whole c-e structure.

The two cases will be formalized by two somewhat different semantics (firing rules): a local and global respectively.

The period of residence of a token at a node is set up on entering this token into it and decreases by one time unit ("tick") of the timer referred to by the node, until permission to leave this node. On expiry of the mandatory residing time at this node, the token can leave it if all other conditions for this action are met. Any elementary c-e structure with the minimal time model can be simulated ("implemented") by an

L. Czaja, *Cause-Effect Structures*, Lecture Notes in Networks and Systems 45, https://doi.org/10.1007/978-3-030-20461-7_4

elementary c-e structure without time but with some additional nodes associated with every original node. A number of these supplementing and linearly ordered nodes represent duration of mandatory stay of a token in the original node.

Definition 4.1.1 (*min-time c-e structure, set* $\boldsymbol{T_{min}CE}$) $U = \langle C, E, T_{min} \rangle$ is a *minimal-time* c-e structure iff $\langle C, E \rangle$ is an elementary c-e structure over $\mathbb{X}$ and T_{min}: $car(U) \to \mathbb{N} \backslash \{0\}$ is a *minimal residence time function* of the informal meaning: $T_{min}(x)$ is the least number of time units indicated by a timer referred to by the node x, during which a token ***must*** stay at x since its appearance. The timer may be either *local*–associated to a particular node, or *global*–shared by all nodes. The set of all the min-time c-e structures over $\mathbb{X}$ is denoted by $\boldsymbol{T_{min}CE}[\mathbb{X}]$. As before, fixing the set of nodes $\mathbb{X}$ let us write $\boldsymbol{T_{min}CE}$ instead of $\boldsymbol{T_{min}CE}[\mathbb{X}]$. □

Definition 4.1.2 (*state*) State is a function $s : \mathbb{X} \to \mathbb{N}$ with the informal meaning: $s(x) = 0$ if there is no token at the node x and $s(x) > 1$ is a remaining time (a number of ticks of the timer referred to by x) during which the token ***must*** remain at x; $s(x) = 1$ indicates that the time of ***compulsory*** residence of a token at the node x, prescribed by $T_{min}(x)$ has elapsed, thus, the token ***can*** be moved further—if other conditions for this are satisfied. The set of all states is $\mathbb{S} = \mathbb{N}^{\mathbb{X}}$ (state-space). □

Remember that now $s(x)$ is not a number of tokens residing at the node x but remaining time of a token's necessary stay at x. This intuitive meaning of state and rules of transition ("firing") between states for local and global minimal time models, are formalized in the two following Definitions 4.1.3 and 4.1.4.

4.1.1 Local Minimal Time

Definition 4.1.3 (*local min-time semantics: a firing rule*) For $Q \in \boldsymbol{FC}[U]$, let $[[Q]]^{loc}_{min} \subseteq \mathbb{S} \times \mathbb{S}$ be a binary relation defined as: $(s, t) \in [[Q]]^{loc}_{min}$ if and only if:

$\forall x \in {}^{\bullet}Q : [s(x) = 1 \wedge t(x) = 0] \wedge \forall y \in Q^{\bullet} : [s(y) = 0 \wedge t(y) = T_{min}(y)] \vee$
$\exists x \in {}^{\bullet}Q^{\bullet} : [s(x) > 1 \wedge t(x) = s(x) - 1]$

Relation $[[Q]]^{loc}_{min}$ expresses passage of the state s to t, performed by the firing component Q.

Semantics $[[U]]^{loc}_{min}$ of $U \in \boldsymbol{T_{min}CE}$ is the union of relations $[[U]]^{loc}_{min} = \bigcup_{Q \in FC[U]} [[Q]]^{loc}_{min}$. Semantics $([[U]]^{loc}_{min})^*$ is the reflexive and transitive closure of $[[U]]^{loc}_{min}$. Parallel min-time semantics $[[U]]^{loc}_{parmin}$ and $([[U]]^{loc}_{parmin})^*$ are defined as $[[U]]^{loc}_{min}$ and $([[U]]^{loc}_{min})^*$ but with $[[Q]]^{loc}_{min}$ replaced with $[[G]]^{loc}_{min}$ where $G \subseteq \boldsymbol{FC}[U]$ is a pairwise detached set of firing components—see Definition 2.8 in Chap. 2. □

The formula $\exists x \in {}^{\bullet}Q^{\bullet} : [s(x) > 1 \wedge t(x) = s(x) - 1]$ expresses decrease by one time unit of token's stay at a certain node x of Q, if the minimal time of this token has not expired in the state s.

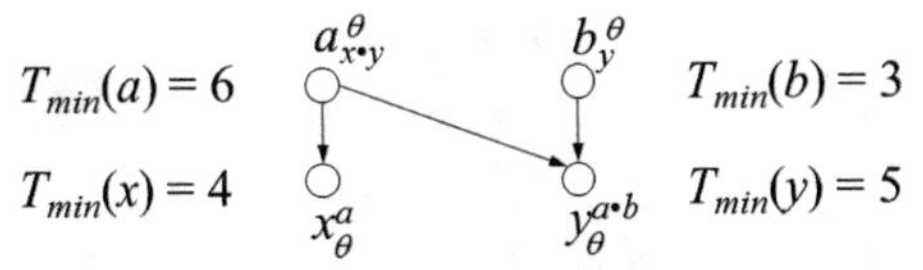

Fig. 4.1 c-e structure $\{a^{\theta}_{x\bullet y}, b^{\theta}_{y}, x^{a}_{\theta}, y^{a\bullet b}_{\theta}\}$ (a single firing component) with min-time assigned to nodes

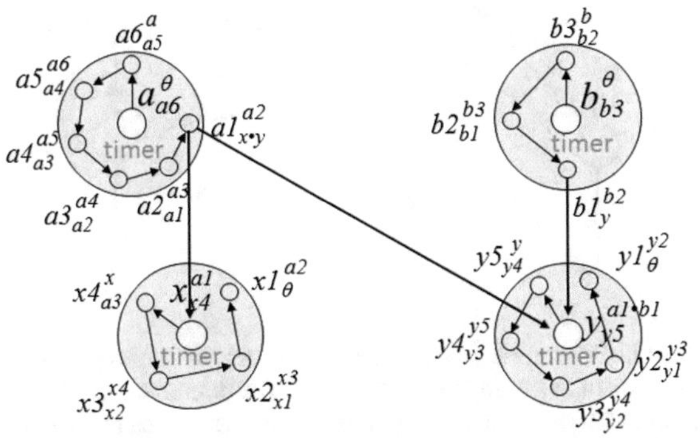

Fig. 4.2 A possible simulation of the c-e structure in Fig. 4.1 with the minimal time of nodes, where $T_{min}(a) = 6$, $T_{min}(b) = 3$, $T_{min}(x) = 4$, $T_{min}(y) = 5$, by means of no-time elementary c-e structure. Here, the separate timer (each with perhaps diverse progress rate of time) is associated with every node. The counterclockwise direction of a token's motion inside the timers, simulates elapse of time controlled by the timers associated with nodes a, b, x, y

Minimal time can be simulated by elementary c-e structures without time constraints. An exemplary simulation of the c-e structure in Fig. 4.1 (firing component) depicts Fig. 4.2.

Still another simulation of the c-e structure in Fig. 4.1 by means of no-time elementary c-e structure, may be accomplished, using a shared timer for any firing component, as shown in Figs. 4.3 and 4.4.

4.1.2 Global Minimal Time

In this variant of c-e structures, the time of their activity is measured by a timer (or oscillator) common to all nodes. The minimal time and state are defined as functions T_{min} and s in Definitions 4.1.1 and 4.1.2, but the semantics is re-interpreted: the decreasing elapse of time now concerns all nodes in $car(U)$, not only a given firing component.

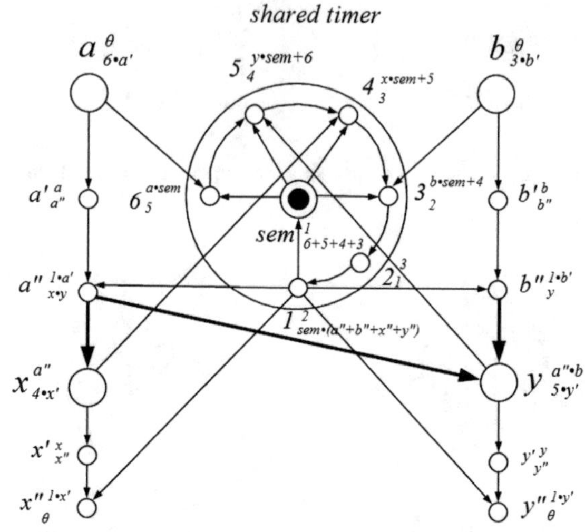

Fig. 4.3 Another simulation of the c-e structure in Fig. 4.1, by means of no-time elementary c-e structure, now with usage of the shared timer. The clockwise direction of a token's motion inside this timer, simulates elapse of minimal mandatory time residence assigned to nodes a, b, x, y. The node sem is a semaphore protecting sequence of nodes 1, 2, 3, 4, 5, 6 inside the timer against presence more than one token in this sequence at a time. The token at sem means that none of the nodes a, b, x, y has started counting its time of mandatory token's holding. Notice however that in this simulation, tokens travel sequentially: none of them may take a move at the same moment with another one—even inside a different node

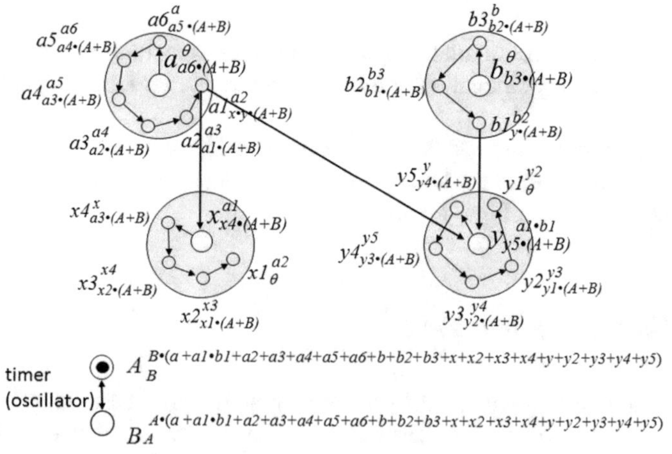

Fig. 4.4 Still another simulation of the c-e structure in Fig. 4.1, now by using an oscillator $A \leftrightarrows B$ to control time flow

Definition 4.1.4 (*global min-time semantics: a firing rule*) For $Q \in \boldsymbol{FC}[U]$, let $[[Q]]^{glob}_{min} \subseteq \mathbb{S} \times \mathbb{S}$ be a binary relation defined as: $(s, t) \in [[Q]]^{glob}_{min}$ if and only if:
$\forall x \in {}^{\bullet}Q : [s(x) = 1 \wedge t(x) = 0] \wedge \forall y \in Q^{\bullet} : [s(y) = 0 \wedge t(y) = T_{min}(y)] \vee$
$\exists x \in car(U) : [s(x) > 1 \wedge t(x) = s(x) - 1]$
Relation $[[Q]]^{glob}_{min}$ and semantics $[[U]]^{glob}_{min}$, $([[U]]^{glob}_{min})*$, $[[U]]^{glob}_{parmin}$, $[[U]]^{*}_{parmin}$ of U are defined as in Definition 4.1.3. □

Notice that the formula $\exists x \in car(U) : [s(x) > 1 \wedge t(x) = s(x) - 1]$ expresses decrease by one "tick" of the **global** (for the whole c-e structure) timer of the remaining time stay of a token at a certain node x in $car(U)$, if the minimal mandatory time of this token stay at x has not expired in the state s.

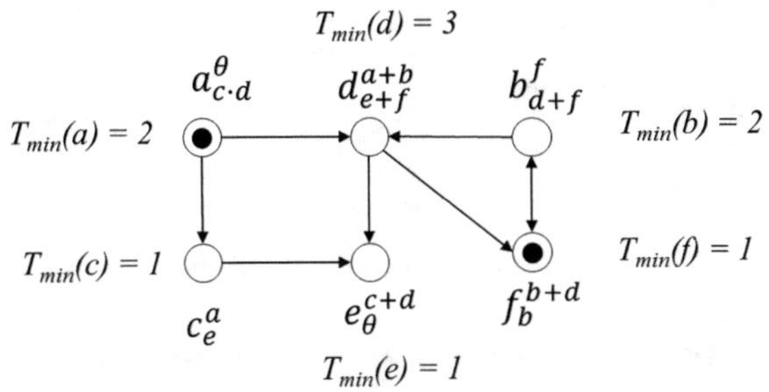

Fig. 4.5 A marked c-e structure with tokens at a and f, and global min-time of nodes

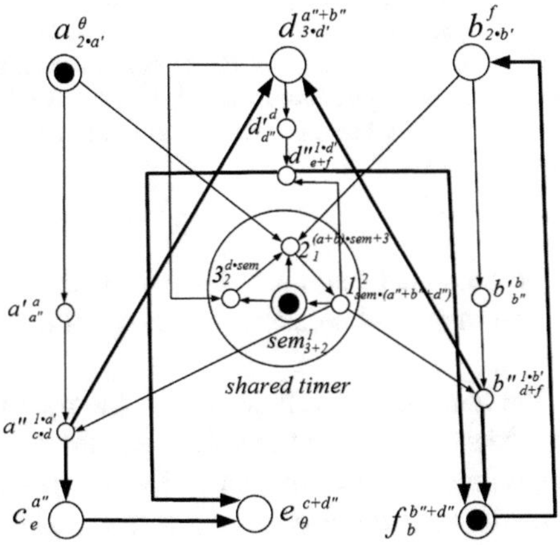

Fig. 4.6 A possible simulation of the c-e structure in Fig. 4.5, with min-time $T_{min}(a) = T_{min}(b) = 2$, $T_{min}(d) = 3$, $T_{min}(c) = T_{min}(e) = T_{min}(f) = 1$, by means of the no-time elementary c-e structure with usage of the shared timer. The clockwise direction of a token's motion inside the shared timer, simulates elapse of minimal time residence assigned to nodes a, b, c, d, e, f and controlled by the timer

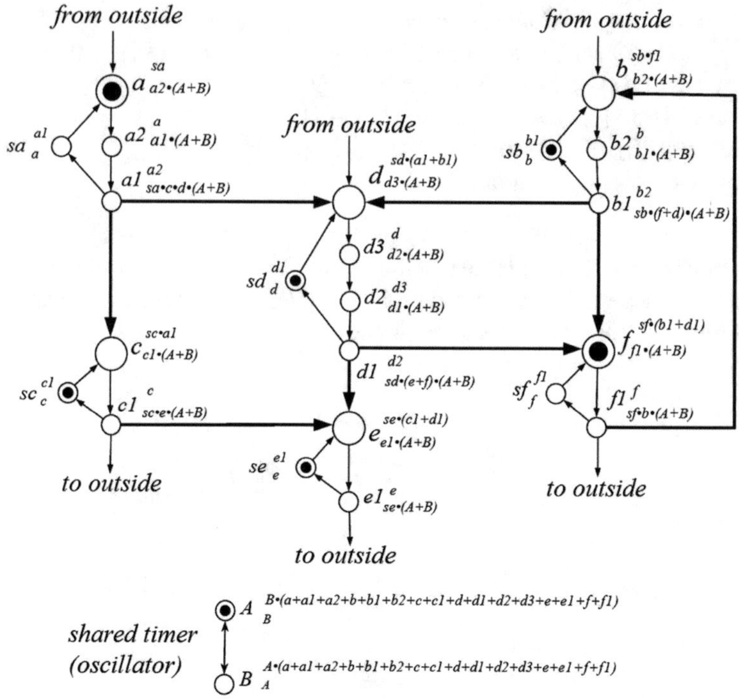

Fig. 4.7 Another possible simulation of the c-e structure in Fig. 4.5 (with the global min-time semantics), by means of the no-time elementary c-e structure and shared oscillator to control the time flow. Nodes sa, sb, sc, sd, se, sf play role of semaphores preventing more than one token within a stream of respective nodes ($a1$, $a2$, a), ($b1$, $b2$, b), etc. which simulate elapse of time. So, absence of token in sa and sf makes its presence in exactly one node of the stream ($a1$, $a2$, a) and ($f1$, f).The marking (distribution of tokens) of this c-e structure corresponds to the marking of global min-time c-e structure in Fig. 4.5

Global minimal time can also be simulated by elementary c-e structures without time constraints. Two exemplary simulations of the c-e structure in Fig. 4.5 depict Figs. 4.6 and 4.7. □

Unlike in case of the local minimal time model, the motion of tokens through nodes is controlled by the single shared timer for the entire c-e structure.

Similarly to the simulation method shown in Fig. 4.4 (using an oscillator as a shared timer), the marked c-e structure in Fig. 4.5 with global time, can be simulated by a no-time c-e structure shown in Fig. 4.7.

Remark A more common model of minimal time, introduced usually for Petri nets, where such time is assigned not to places but to transitions, may be obtained for c-e structures as a special case of the one introduced in Definition 4.1.1, assuming that a firing component Q can fire if the delay time of the longest time $T_{min}(x)$ for all preconditions $x \in {}^{\bullet}Q$ has elapsed since enabling of Q and if a joint oscillator to measure time of all its nodes is applied. An exemplary simulation of a firing

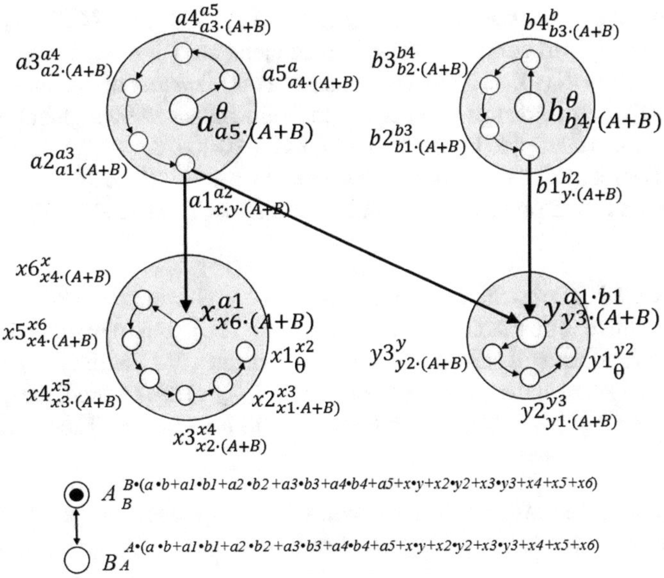

Fig. 4.8 A possible simulation of the local-min-time firing component $Q = \{a^{\theta}_{x\bullet y}, b^{\theta}_{y}, x^{a}_{\theta}, y^{a\bullet b}_{\theta}\}$ with $T_{min}(a) = 5$, $T_{min}(b) = 4$, $T_{min}(x) = 6$, $T_{min}(y) = 3$ by means of no-time c-e structure. Notice that the streams of arrows $a \rightarrow a5 \rightarrow ... \rightarrow a1$ and $b \rightarrow b4 \rightarrow ... \rightarrow b1$ are being traversed by tokens synchronously (simultaneously), and the same concerns the streams $x \rightarrow x6 \rightarrow ... \rightarrow x1$ and $y \rightarrow y3 \rightarrow y2 \rightarrow y1$. This is ensured by polynomials in the oscillator $A \leftrightarrows B$

component with a minimal delay time by no-time elementary c-e structure is in Fig. 4.8. Notice that some special time models (such as e.g. *time* or *timed* Petri nets), though do not discussed here, can be incorporated and respective c-e structures can be simulated by suitable selection of polynomials in the timer.

4.2 Maximal Time Model

In the maximal time model, with each node of every firing component, a time period of ***permissible*** stay of a token at this node is associated. This is the longest time during which the node ***can*** hold the token. On expiry of this period, the token ***must*** leave the node as soon as possible (provided that other necessary conditions for this "move" are met) prior to any move of tokens with unexpired time, residing in other firing components. Therefore, in contrast to the minimal-time model, this model requires to scrutinize the state of all firing components of the c-e structure, i.e. to make use of a global information. The time flow is measured by a timer shared by all nodes of the c-e structure.

Definition 4.2.1 (*max-time c-e structure, set $\boldsymbol{T_{max}CE}$*) $U = \langle C, E, T_{max}\rangle$ is a *maximal-time* c-e structure iff $\langle C, E\rangle$ is an elementary c-e structure over $\mathbb{X}$ and T_{max}: $car(U) \to \mathbb{N}\backslash\{0\}$ is a *maximal residence time function* of the informal meaning: $T_{max}(x)$ is the greatest number of time units indicated by a global (shared by all nodes) timer referred to by the node x, thus the period, during which a token ***can*** stay at x. The set of all the max-time c-e structures over $\mathbb{X}$ is denoted by $\boldsymbol{T_{max}CE}[\mathbb{X}]$. As before, fixing the set of nodes $\mathbb{X}$ let us write $\boldsymbol{T_{max}CE}$ instead of $\boldsymbol{T_{max}CE}[\mathbb{X}]$.

□

Definition 4.2.2 (*state*) State is a function $s : \mathbb{X} \to \mathbb{N}$ with the informal meaning: $s(x) = 0$ if there is no token at the node x and $s(x) > 1$ is a remaining time during which the token ***can*** still stay at x; $s(x) = 1$ indicates that the time of ***permissible*** residence of a token at the node x, prescribed by $T_{max}(x)$ has expired, thus, the token ***must*** be moved further—if other conditions for this are satisfied. The set of all states is $\mathbb{S} = \mathbb{N}^{\mathbb{X}}$ (state-space). □

Before defining the maximal time semantics, let us introduce two predicates Φ and Ψ characterizing conditions for a given firing component $Q \in \boldsymbol{FC}[U]$ to transform a state s into t.

The formula $\Phi(Q, s, t)$ expresses what follows:

In the state s, every node in $^\bullet Q$ holds a token, but none in $Q^\bullet$ does, and the max-time stay of a token at a certain node in $^\bullet Q$—has expired; in the next state t, no node in $^\bullet Q$ holds a token, but all nodes in $Q^\bullet$ do, and their tokens can stay there for maximal time; moreover, if in the state s, at whichever node outside Q, is a token with unexpired time, then in the next state t, the remaining time of this token to stay at this node, is decreased by one unit.

In symbols: $\Phi(Q, s, t)$ if and only if:

$$\begin{aligned}
&\forall x \in {}^\bullet Q : s(x) > 0 \wedge t(x) = 0 \wedge \exists x \in {}^\bullet Q : s(x) = 1 \wedge\\
&\forall x \in Q^\bullet : s(x) = 0 \wedge t(x) = T_{max}(x) \wedge\\
&\exists x \notin {}^\bullet Q^\bullet : s(x) > 1 \wedge t(x) = s(x) - 1
\end{aligned}$$

The formula $\Psi(Q, s, t)$ expresses what follows:

In the state s, every node in $^\bullet Q$ holds a token, with unexpired time, but none in $Q^\bullet$ holds a token; in the next state t, no node in $^\bullet Q$ holds a token, but all nodes in $Q^\bullet$ do, and their tokens can stay there for maximal time; besides, no firing component $P \in \boldsymbol{FC}[U]$, different from Q, satisfies $\Phi(P, s, t)$, i.e. forced to fire; moreover, if in the state s, at whichever node outside Q, is a token with unexpired time, then in the next state t, the remaining time of this token to stay at this node, is decreased by one unit.

In symbols: $\Psi(Q, s, t)$ if and only if:

$$\begin{aligned}
&\forall x \in {}^\bullet Q : s(x) > 1 \wedge t(x) = 0 \wedge\\
&\forall x \in Q^\bullet : s(x) = 0 \wedge t(x) = T_{max}(x) \wedge\\
&\neg\exists P \in \boldsymbol{FC}[U] : \Phi(P, s, t) \vee\\
&\exists x \in car(U) : s(x) > 1 \wedge t(x) = s(x) - 1
\end{aligned}$$

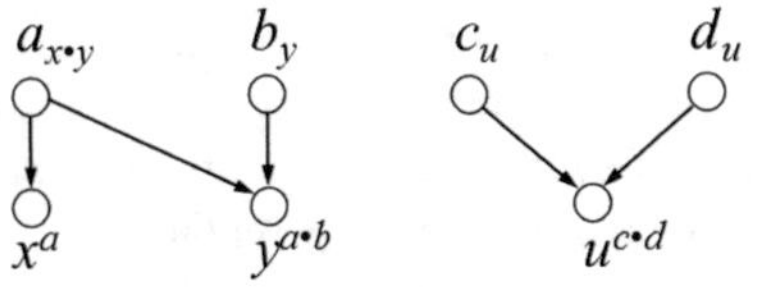

Fig. 4.9 A max-time c-e structure with $T_{max}(a) = 5,\ T_{max}(b) = 4,\ T_{max}(x) = 6,\ T_{max}(y) = 3,$ $T_{max}(c) = 3,\ T_{max}(d) = 5,\ T_{max}(u) = 4$

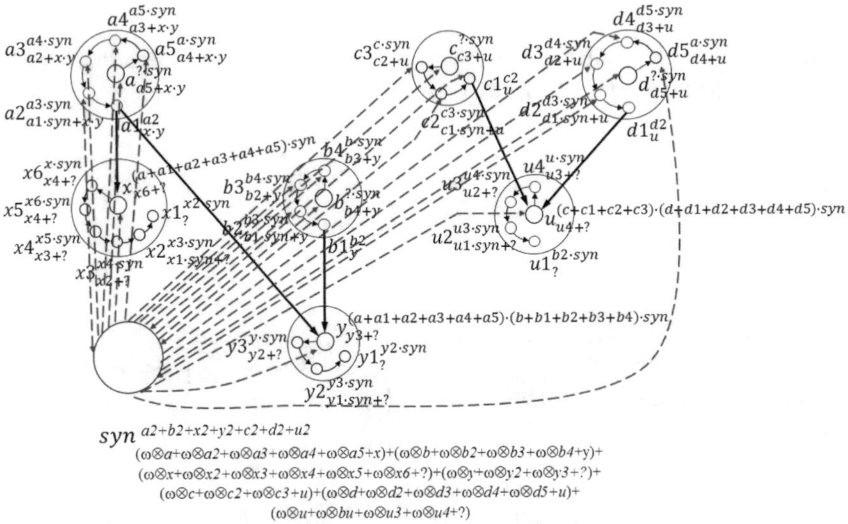

Fig. 4.10 A possible simulation of the two firing components c-e structure in Fig. 4.9, by no-time c-e structure with added synchroniser *syn*—the inhibitor node, from which protrude dashed red inhibiting arrows

Definition 4.2.3 (*max-time semantics: a firing rule*) For $Q \in \boldsymbol{FC}[U]$, let $[[Q]]_{max} \subseteq \mathbb{S} \times \mathbb{S}$ be a binary relation defined as: $(s, t) \in [[Q]]_{max}$ if and only if $\Phi(Q, s, t) \vee \Psi(Q, s, t)$. Relation $[[Q]]_{max}$ expresses transition of state s to t performed by the firing component Q. As before, semantics $[[U]]_{max}$ of $U \in \boldsymbol{T}_{max}\boldsymbol{CE}$ is the union of relations $[[U]]_{max} = \bigcup_{Q \in FC[U]} [[Q]]_{max}$. Also the relations $([[U]]_{max})^*$, $[[U]]_{parmax}$ and $([[U]]_{parmax})^*$ are defined as before.

□

Likewise the minimal time of the elementary c-e structures, the maximal time can be simulated by c-e structures without time, but using some added inhibitor nodes. An example is of the simulation of max-time c-e structure (containing only two firing components) depicted in Fig. 4.9, is a no-time c-e structure with the inhibitor node *syn*, depicted in Fig. 4.10.

Notice that the superscript polynomials $C(y)$ and $C(u)$ in Fig. 4.10 are written in a concise form, instead of the canonical form (see the last Remark on Definition 2.1)—as sums of products of every symbol contained in one pair of parentheses by every symbol contained in the other pair of parentheses.

4.2.1 Example (Music Text)

Figure 4.11 shows a simulation of a global min-time c-e structure presenting a music text (score), by no-time c-e structure.

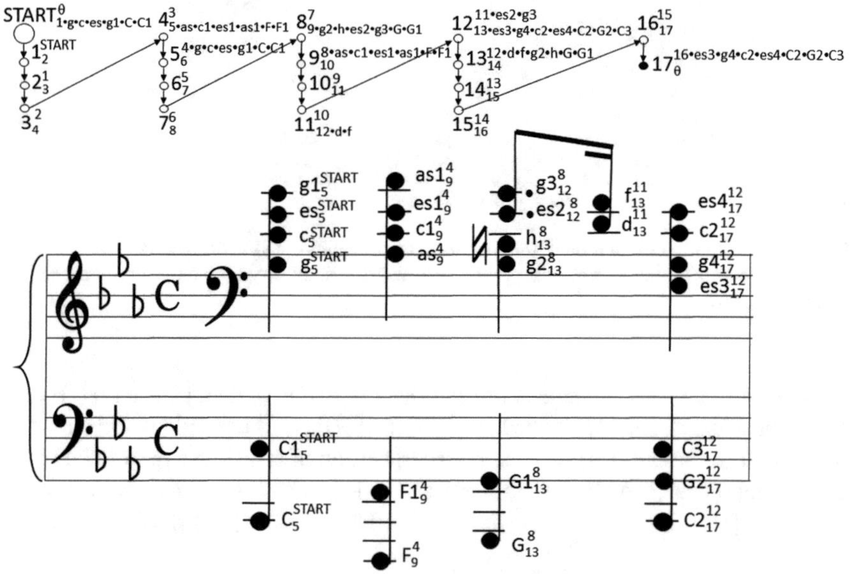

Fig. 4.11 First bar of the score of Prelude c-minor by Chopin, in the form of the min-time c-e-structure simulated by a no-time c-e structure. The notes are represented as nodes with assigned duration periods, which are implemented by the control mechanism above the music text. The chords are accomplished by synchronisation of notes lying on one vertical line; synchronising—by means of multiplication "•"

Chapter 5
Rough Cause-Effect Structures

Usage of the c-e structures as objects of some information systems has been inspired by the Rough Set (RS) Theory, introduced by Pawlak [3], then being developed by many authors (for a few publications—see References [4–9]). The idea behind the RS is a characterization of given objects in terms of their features essential for solving some tasks, for making decisions in particular. The objects of identical features, thus indistinguishable for the viewpoint of a task, are collected into a group, and family of such groups is a partition of the set of all objects in this task. Usually not every subset of all objects may be unambiguously characterized by intended features, thus the "approximations" of such subsets are introduced: lower and upper. Such subsets are called *rough* relative to the features. The idea can be applied to c-e structures. Here, the information system, in adaptation to them is defined as follows.

Definition 5.1 (*information system of c-e structures*) An information system of c-e structures is a triple $\langle \mathbb{U}, \mathbb{A}, \{\sigma_a\}_{a \in \mathbb{A}} \rangle$, where:

- $\mathbb{U} \subseteq \boldsymbol{CE}[\mathbb{X}]$ is a set of objects, here c-e structures over $\mathbb{X}$
- $\mathbb{A}$ is a set of attributes (features of members of $\mathbb{U}$), with each attribute $a \in \mathbb{A}$ is associated its set of values V_a
- $\{\sigma_a\}_{a \in \mathbb{A}}$ is a set of functions $\sigma_a : \mathbb{U} \to V_a$, called *mappings of similarity* □

Similarity is a relative property: some c-e structures may be similar with respect to some features, whereas not to others. So, a given subset $\mathbb{B} \subseteq \mathbb{A}$ defines groups of similar c-e structures: $U_1, U_2 \in \mathbb{U}$ are similar relative to $\mathbb{B}$ when $\sigma_a(U_1) = \sigma_a(U_2)$ for all $a \in \mathbb{B}$. Say then: "U_1, U_2 are $\mathbb{B}$-similar".

Definition 5.1 is illustrated by example of the so-called "Revised-Review" problem (originated from [1]) in terms of c-e structures. The original story is told here as a process of editorial elaboration of books by persons working for a publishing company.

L. Czaja, *Cause-Effect Structures*, Lecture Notes in Networks and Systems 45, https://doi.org/10.1007/978-3-030-20461-7_5

5.1 Example (Revise/Review of a Book)

This task consists in revision (proof-reading), then more thorough review of books to be published, and may proceed cyclically. There are n books to be revised, then to be essentially reviewed by m persons. Every person can serve as a revisor as well as reviewer (perhaps for a shortage of personnel). Initially, every book is neither revised nor reviewed and the persons are idle. The action cycle in which a book is being revised, then reviewed is the following:

- a currently idle person takes the book from a set for **revision**, then after having revised, puts it into a set for **review** and becomes idle.
- a currently idle person takes the book from a set for **review**, then after having reviewed, puts it into a set for **revision** and becomes idle.

The revise/review, being made by the publisher, is specified by the c-e structure in Fig. 5.1, depicted in the bus layout (see Fig. 2.4 in Chap. 2) with $n = 4$ books and $m = 3$ persons. Meaning of nodes is the following:

- **1**, **2**, **3** holds a token if the person numbered correspondingly $1, 2, 3,$ is idle.The numerals in the bold font are names of nodes representing persons, whereas the corresponding digits in the normal font, are indices in composite names, like "$r[j, k]$", where $j = 1, 2, 3, 4$ and $k = 1, 2, 3,$
- $a[j]$ holds a token if the book with number j can be taken from the set {book 1, book 2, book 3, book 4} for **revision** by an idle person,
- $r[j, k]$ holds a token if the book with number j is being **revised** by the person number k,
- $b[j]$ holds a token if the book with number j, after revision, can be taken from the set {book 1, book 2, book 3, book 4} for **review** by an idle person,
- $R[j, k]$ holds a token if the book with number j is being **reviewed** by the person of number k.

Initially, only nodes $a[1], a[2], a[3], a[4]$ and **1**, **2**, **3** hold tokens.

In Fig. 5.1 with $n = 4$ and $m = 3$, the upper and lower polynomials of nodes are omitted. The annotated nodes, though readable from the Figure, are the following:

$$\mathbf{1}^{R[1,1]+r[1,1]+R[2,1]+r[2,1]+R[3,1]+r[3,1]+R[4,1]}_{R[1,1]+r[1,1]+R[2,1]+r[2,1]+R[3,1]+r[3,1]+r[4,1]},$$

$$\mathbf{2}^{R[1,2]+r[1,2]+R[2,2]+r[2,2]+R[3,2]+r[3,2]+R[4,2]}_{R[1,2]+r[1,2]+R[2,2]+r[2,2]+R[3,2]+r[3,2]+r[4,2]},$$

$$\mathbf{3}^{R[1,3]+r[1,3]+R[2,3]+r[2,3]+R[3,3]+r[3,3]+R[4,3]}_{R[1,3]+r[1,3]+R[2,3]+r[2,3]+R[3,3]+r[3,3]+r[4,3]},$$

$$a[1]^{R[1,1]+R[1,2]+R[1,3]}_{r[1,1]+r[1,2]+r[1,3]},\quad b[1]^{r[1,1]+r[1,2]+r[1,3]}_{R[1,1]+R[1,2]+R[1,3]},$$

$$r[1,1]^{a[1]\bullet\mathbf{1}}_{b[1]\bullet\mathbf{1}},\quad r[1,2]^{a[1]\bullet\mathbf{2}}_{b[1]\bullet\mathbf{2}},\quad r[1,3]^{a[1]\bullet\mathbf{3}}_{b[1]\bullet\mathbf{3}},$$

$$R[1,1]^{b[1]\bullet\mathbf{1}}_{a[1]\bullet\mathbf{1}},\quad R[1,2]^{b[1]\bullet\mathbf{2}}_{a[1]\bullet\mathbf{2}},\quad R[1,3]^{b[1]\bullet\mathbf{3}}_{a[1]\bullet\mathbf{3}},$$

$$a[2]^{R[2,1]+R[2,2]+R[2,3]}_{r[2,1]+r[2,2]+r[2,3]},\quad b[2]^{r[2,1]+r[2,2]+r[2,3]}_{R[2,1]+R[2,2]+R[2,3]},$$

$$r[2,1]^{a[2]\bullet 1}_{b[2]\bullet 1},\ r[2,2]^{a[2]\bullet 2}_{b[2]\bullet 2},\ r[2,3]^{a[2]\bullet 3}_{b[2]\bullet 3},$$

$$R[2,1]^{b[2]\bullet 1}_{a[2]\bullet 1},\ R[2,2]^{b[2]\bullet 2}_{a[2]\bullet 2},\ R[2,3]^{b[2]\bullet 3}_{a[2]\bullet 3},$$

$$a[3]^{R[3,1]+R[3,2]+R[3,3]}_{r[3,1]+r[3,2]+r[3,3]},\ b[3]^{r[3,1]+r[3,2]+r[3,3]}_{R[3,1]+R[3,2]+R[3,3]},$$

$$r[3,1]^{a[3]\bullet 1}_{b[3]\bullet 1},\ r[3,2]^{a[3]\bullet 2}_{b[3]\bullet 2},\ r[3,3]^{a[3]\bullet 3}_{b[3]\bullet 3},$$

$$R[3,1]^{b[3]\bullet 1}_{a[3]\bullet 1},\ R[3,2]^{b[3]\bullet 2}_{a[3]\bullet 2},\ R[3,3]^{b[3]\bullet 3}_{a[3]\bullet 3},$$

$$a[4]^{R[4,1]+R[4,2]+R[4,3]}_{r[4,1]+r[4,2]+r[4,3]},\ b[4]^{r[4,1]+r[4,2]+r[4,3]}_{R[4,1]+R[4,2]+R[4,3]},$$

$$r[4,1]^{a[4]\bullet 1}_{b[3]\bullet 1},\ r[4,2]^{a[4]\bullet 2}_{b[3]\bullet 2},\ r[4,3]^{a[4]\bullet 3}_{b[3]\bullet 3},$$

$$R[4,1]^{b[4]\bullet 1}_{a[4]\bullet 1},\ R[4,2]^{b[4]\bullet 2}_{a[4]\bullet 2},\ R[4,3]^{b[4]\bullet 3}_{a[4]\bullet 3}$$

The set of these annotated nodes makes the c-e structure, presented in graphic form by Fig. 5.1.

Components of the information system $\langle \mathbb{U}, \mathbb{A}, \{\sigma_a\}_{a\in\mathbb{A}}\rangle$ in Definition 5.1, for the revise/review task are the following:

- $\mathbb{U} =$ {book 1, book 2, book 3, book 4} –the set of objects: c-e structures depicted in Fig. 5.2,
- $\mathbb{A} =$ {*documented*, *illustrated*, linguistic *correctness*, *size*, *innovative*, *publishable*}—the set of attributes related to the objects given in Table 5.1
- Functions $\sigma_a : \mathbb{U} \rightarrow V_a$, where $a \in \mathbb{A}$ and $V_a =$ {good, medium, no, poor, rich, small, well, yes} with their values are given in Table 5.1.

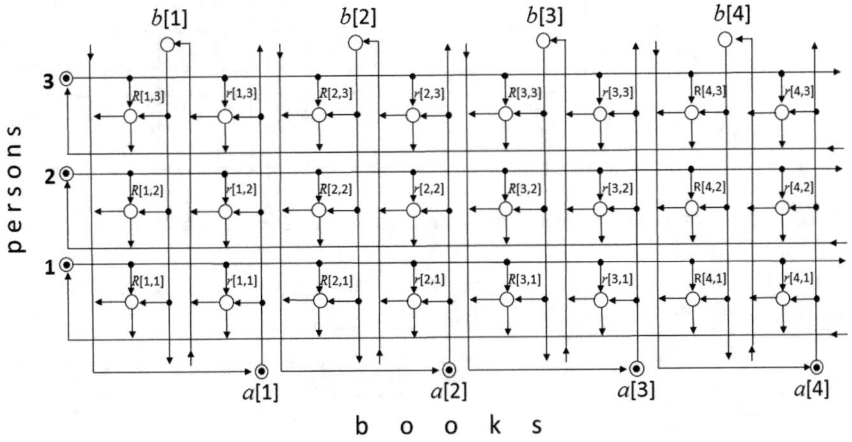

Fig. 5.1 Revise/review system of 4 books and 3 working persons in initial state; the "bus layout" allows for direct visual extension onto any number of books and persons

Assume that the subset of attributes most important for the publisher is: $\mathbb{B} = \{$*illustrated*, *l inguistic correctness*, *size*$\} \subset \mathbb{A}$.

Thus, book 1, book 2, book 3 are similar (indistinguishable) with respect to $\mathbb{B}$, because σ_a(book 1) $= \sigma_a$(book 2) $= \sigma_a$(book 3) for every $a \in \mathbb{B}$.

It is evident that the similarity is an equivalence relation, thus partitions the set $\mathbb{U}$ onto two similarity classes {book 1, book 2, book 3}, book 4}. The books in the same class are called $\mathbb{B}$-similar.

From the Table 5.1 one may expect that books 1 and 4 will be published, so, they make the so-called target set $\mathbb{T} = \{$book 1, book 4$\} \subset \mathbb{U}$. The books are represented by c-e structures (Fig. 5.2), so we define the target c-e structure as the sum: book 1 + book 4, depicted in Fig. 5.3 and being the substructure of that in Fig. 5.1.

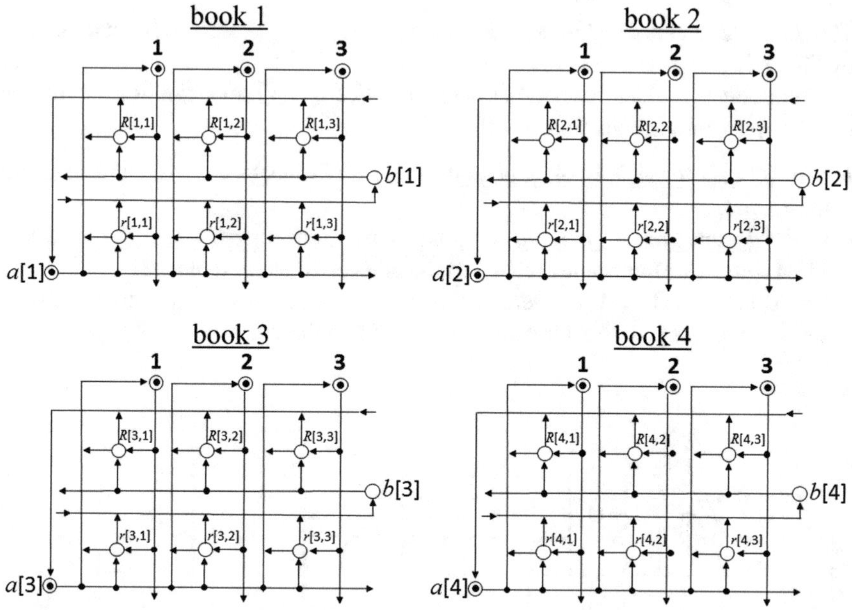

Fig. 5.2 Objects of the information system—substructures of c-e structure in Fig. 5.1

Table 5.1 Attributes of objects, their values and the mappings $\{\sigma_a\}_{a \in \mathbb{A}}$. For instance, σ_{size} (book 1) = medium, $\sigma_{innovative}$ (book 3) = no.

Object	Attribute					
	documented	illustrated	linguistic correctness	size	innovative	publishable
book 1	well	rich	good	medium	yes	yes
book 2	no	rich	good	medium	poor	no
book 3	poor	rich	good	medium	no	no
book 4	yes	yes	medium	small	medium	yes

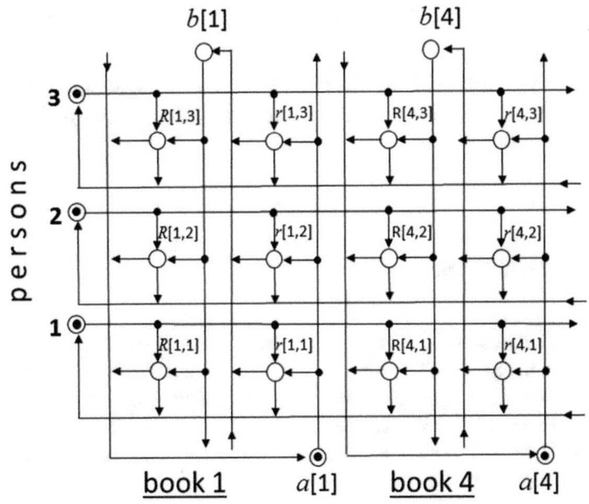

Fig. 5.3 Target c-e structure book 1 + book 4

Since book 1 is in $\mathbb{T}$, but some $\mathbb{B}$-similar to it are not, the exact representation by attributes from $\mathbb{B}$ of all publishable books is impossible. Hence some approximate representations are introduced.

The approximations (relative to $\mathbb{B}$) of the target $\mathbb{T}$ are defined as follows:

The *lower approximation* is a set $\underline{\mathbb{B}}\mathbb{T}$ of books, which are in $\mathbb{T}$ along with *all* $\mathbb{B}$-similar to them, so $\underline{\mathbb{B}}\mathbb{T} = \{$book 4$\}$ (every book is $\mathbb{B}$-similar to itself).

The *upper approximation* is a set $\overline{\mathbb{B}}\mathbb{T}$ of books, whose *some* $\mathbb{B}$-similar to them are in $\mathbb{T}$, so $\overline{\mathbb{B}}\mathbb{T} = \{$book 1, book 2, book 3, book 4$\}$. Thus, respective c-e structure is their sum $\sum \overline{\mathbb{B}}\mathbb{T}$ depicted in Fig. 5.1.

In accordance with the Rough Set Theory, the approximations and boundary of a target set, in case of the c-e structures as objects, are defined as follows:

Definition 5.2 (*approximations, boundary, rough c-e structure*) Given an information system of rough c-e structures $\langle \mathbb{U}, \mathbb{A}, \{\sigma_a\}_{a \in \mathbb{A}} \rangle$, let $\mathbb{B} \subseteq \mathbb{A}$, $\mathbb{T} \subseteq \mathbb{U}$, $U \in \mathbb{U}$ and let $[U]_B$ denote the class of c-e structures similar to U with respect to $\mathbb{B}$. Then:

$\underline{\mathbb{B}}\mathbb{T} = \{U \in \mathbb{U} : [U]_{\mathbb{B}} \subseteq \mathbb{T}\}$ is the *lower approximation* of $\mathbb{T}$ relative to $\mathbb{B}$.

$\overline{\mathbb{B}}\mathbb{T} = \{U \in \mathbb{U} : [U]_B \cap \mathbb{T} \neq \emptyset\}$ is the *upper approximation* of $\mathbb{T}$ relative to $\mathbb{B}$.

$\overline{\mathbb{B}}\mathbb{T} \setminus \underline{\mathbb{B}}\mathbb{T}$ is the *boundary region* of $\mathbb{T}$ relative to $\mathbb{B}$.

The c-e structure $\sum \mathbb{T}$ is rough relative to $\mathbb{B}$ when $\sum \overline{\mathbb{B}}\mathbb{T} \neq \sum \underline{\mathbb{B}}\mathbb{T}$. □

Obviously $\sum \underline{\mathbb{B}}\mathbb{T} \le \sum \mathbb{T} \le \sum \overline{\mathbb{B}}\mathbb{T} \le \sum \mathbb{U}$ (for "≤"—see Definition 2.4).

The RS theory introduces a number of accuracy measures for approximations, the simplest is the quotient $\frac{|\underline{\mathbb{B}}\mathbb{T}|}{|\overline{\mathbb{B}}\mathbb{T}|}$, thus in case of the revise/review above example it is $\frac{1}{4}$. A choice of a measure, relevant to a given task, depends on the task body (content). For instance one may consider the quotient $\frac{\text{size of } \sum \underline{\mathbb{B}}\mathbb{T}}{\text{size of } \sum \overline{\mathbb{B}}\mathbb{T}}$ where size of a c-e structure is e.g. number of its nodes. Some other measures might be considered—cf.

[2]. Likewise, such concepts of the RS theory as the *reduct* and *refinement* (adding some new attributes but retaining the old), can be adopted to the information system of c-e structures. But this chapter is mere an example, that some tasks with approximate outcomes and specified in terms of c-e structures, might be approached using concepts of the RS theory.

References

1. Henderson P (1989) A talk given at the IFIP W.G.2.3 24th meeting in Zaborow, Poland
2. Nguyen HS, Skowron A (2013) Rough sets: from rudiments to challenges. In Skowron A, Suraj Z (eds) Rough sets and intelligent systems, ISRL 42. Springer, Berlin Heidelberg, pp 75–173
3. Pawlak Z (1982) Rough Sets. Int J Comput Inf Sci 11(5):341–356
4. Pawlak Z (1991) Rough sets: theoretical aspects of reasoning about data. Kluwer Academic Publishers, Dordrecht
5. Pawlak Z (1992) Concurrent versus sequential the rough sets perspective. Bull EATCS 48:178–190
6. Pawlak Z (2002) Rough sets theory and its applications. J Telecommun Inf Technol 3:7–10
7. Peters JF, Skowron A, Suraj Z, Ramanna S (1999) Guarded transitions in rough Petri nets. In Proceedings of 7th European Congress on Intelligent Systems & Soft Computing (EUFIT'99), 13–16 Sept 1999. Aachen, Germany, pp 2003–2012
8. Skowron A, Suraj Z (1993) Rough sets and concurrency. Bull Pol Acad Sci 41–3:237–254
9. Suraj Z (2000) Rough set methods for the synthesis and analysis of concurrent processes. In Polkowski L, et al.(eds) Rough set methods and applications. Springer, Berlin, pp 379–488

Chapter 6
Structural Properties—Lattice of c-e Structures

In this chapter a number of structural properties of c-e structures are stated and proved. Since these properties concern the structure only, not the behaviour, they pertain to all kinds of c-e structures considered so far. Some properties are direct consequences of equations in Proposition 2.1 defining the quasi-semiring, and have been listed in Proposition 2.2. Some, more mathematically involved, follow from general facts in the lattice theory, since, as Theorems 6.1 and 6.2 state, the set $\boldsymbol{CE}[\mathbb{X}]$ of all c-e structures and the set $\boldsymbol{SUB}[U]$ of all substructures of a c-e structure U, when partially ordered by "$\leq$" (Definition 2.4), become lattices. From the practical point of view, the properties appeared helpful in composing large c-e structures from small components, as well as in simplification of c-e structures, that is, transformation leading to simpler but behaviourally equivalent system descriptions. Remember that the set $\mathbb{X}$ of nodes, out of which the c-e structures are built, is assumed infinite, and any c-e structure contains all its members, also the isolated—without predecessors and successors. So, each c-e structure is, per se, infinite, but its carrier (Definition 2.2) may be either finite or infinite. Since we attempt to combine c-e structures also with infinite carriers (examples are processes—the unfoldings of c-e structures—Chaps. 10, 11, 12), some infinite operations on them will be needed.

6.1 Structural Properties of the Least Upper Bound of Sets of c-e Structures

For a set of c-e structures $A \subseteq \boldsymbol{CE}[\mathbb{X}]$, the least upper bound of A, denoted by $\sum A$, is the least (wrt $\leq$) c-e structure V with $U \leq V$ for each $U \in A$. In addition, let $\sum \emptyset = \theta$. If I is a certain index set then, as customarily, denote: $\sum\limits_{j \in I} U_j = \sum\{U_j |\ j \in I\}$ or $\sum\limits_{j=0}^{n} U_j$ when $I = \{0, 1, ..., n\}$ or $\sum\limits_{j=0}^{\infty} U_j$ or $\sum\limits_{j \geq 0} U_j$ when $I = \mathbb{N}$. Obviously, for $U_j \in$

L. Czaja, *Cause-Effect Structures*, Lecture Notes in Networks and Systems 45, https://doi.org/10.1007/978-3-030-20461-7_6

$\boldsymbol{CE}[\mathbb{X}]$: $\sum\{U_1, U_2, ..., U_n\} = U_1 + U_2 + ... + U_n \in \boldsymbol{CE}\ [\mathbb{X}]$. However for infinite A, not always $\sum A$ exists. For instance, let $A = \{\{a_1, \mathbf{1}^a\}, \{a_2, \mathbf{2}^a\}, \{a_3, \mathbf{3}^a\}, ...\}$. Here, the bold numerals are names of nodes. Obviously, $A \subseteq \boldsymbol{CE}\ [\mathbb{X}]$. Suppose $\sum A \in \boldsymbol{CE}[\mathbb{X}]$ and let $U_i = \{a_{\mathbf{i}}, \mathbf{i}^a\}(i = 1, 2, 3, ..., \quad \mathbf{i} = \mathbf{1}, \mathbf{2}, \mathbf{3}, ...)$ and $S_i = U_1 + U_2 + ... + U_i$. Hence, the subscript polynomial of a in S_{i+1} is longer than that in S_i. Since each $S_i \leq \sum A$ then the subscript polynomial of a in $\sum A$ would be of infinite length: $\mathbf{1} + \mathbf{2} + \mathbf{3} + ...$ thus, $\sum A \notin \boldsymbol{CE}[\mathbb{X}]$—a contradiction. There exist, however, c-e structures with unbounded length of polynomials. An example is:

$$U = \{\mathbf{1}_{\mathbf{2}}, \mathbf{2}^{\mathbf{1}}_{\mathbf{3+4}}, \mathbf{3}^{\mathbf{2}}_{\mathbf{4+5+6}}, \mathbf{4}^{\mathbf{3+2}}_{\mathbf{5+6+7+8}}, \mathbf{5}^{\mathbf{4+3}}_{\mathbf{6+7+8+9+10}}, \mathbf{6}^{\mathbf{5+4+3}}_{\mathbf{7+8+9+10+11+12}}, ...\}$$

This c-e structure is depicted in Fig. 6.1. Obviously, it is the least upper bound of the set $\{U_1, U_2, U_3, ...\}$ where $U_j = \sum_{i=j+1}^{2\cdot j} \{\mathbf{j}^{\theta}_{\mathbf{i}}, \mathbf{i}^{\mathbf{j}}_{\theta}\}$, that is, $U = \sum_{j\geq 1} U_j$; (numbers j and i correspond to nodes $\mathbf{j}, \mathbf{i}$).

Some properties of the least upper bounds of sets of substructures of a given $U \in \boldsymbol{CE}$ (let us omit $\mathbb{X}$) are in:

Proposition 6.1 *(a) For each* $A \subseteq \boldsymbol{CE}$, *if* $\sum A$ *exists then* $A \subseteq \boldsymbol{SUB}[\sum A]$
(b) For each $U \in \boldsymbol{CE}$ *there exists* $\sum \boldsymbol{SUB}[U]$ *and* $\sum \boldsymbol{SUB}[U] = U$
(c) $\bigcup_{j\in I} \boldsymbol{SUB}[U_j] \subseteq \boldsymbol{SUB}[\sum_{j\in I} U_j]$
(d) If $A \subseteq \boldsymbol{SUB}[U]$ *then there exists* $\sum A$ *in* $\boldsymbol{SUB}[U]$

Proof **Of (a)** $U \in A \Rightarrow U \leq \sum A \Leftrightarrow U \in \boldsymbol{SUB}[\sum A]$.

Of (b) Let $V \in \boldsymbol{CE}$ satisfy $W \leq V$ for each $W \in \boldsymbol{SUB}[U]$. There exists indeed such V since $U \in \boldsymbol{SUB}[U]$ and $W \leq U$. Thus $U \leq V$, which means that U is the least of such V's. Hence $U = \sum \boldsymbol{SUB}[U]$.

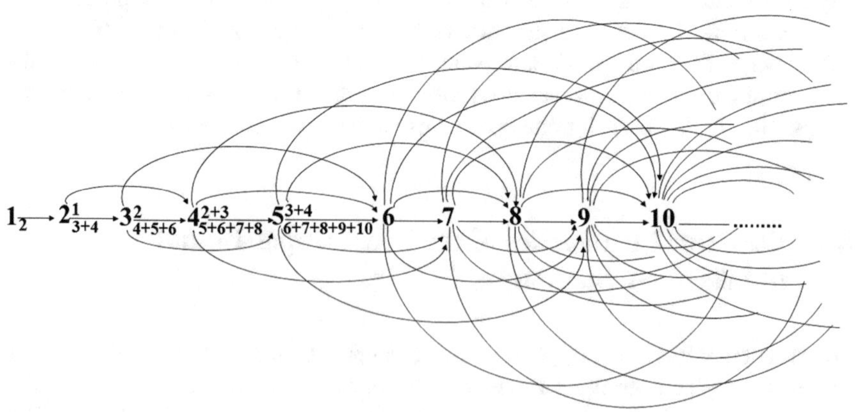

Fig. 6.1 Infinite c-e structure with polynomials of unbounded length

Of (c) Let $U \in \bigcup_{j \in I} \boldsymbol{SUB}[U_j]$. Thus, for a certain $k \in I : U \in \boldsymbol{SUB}[U_k]$. Since $U_k \leq \sum_{j \in I} U_j$ and $U \leq U_k$, we get $U \leq \sum_{j \in I} U_j$, hence $U \in \boldsymbol{SUB}[\sum_{j \in I} U_j]$.

Of (d) Consider a node $x \in \mathbb{X}$, so, $C_U(x)$ is a polynomial, thus, an expression of a finite length. Since for each $V \in A$, polynomial $C_V(x)$ is a sub-polynomial of $C_U(x)$ (meaning $C_U(x) = C_U(x) + C_V(x)$), then there are finitely many V's in A, whose polynomials are pairwise distinct. The sum of these distinct polynomials equals $\sum_{V \in A} C_V(x)$. Analogously, $\sum_{V \in A} E_V(x)$ is obtained from $E_U(x)$. Define $W = (C_W, E_W)$ with $C_W(x) = \sum_{V \in A} C_V(x)$, $E_W(x) = \sum_{V \in A} E_V(x)$. Easy checking ensures that:

(i) W is a c-e structure (i.e. y occurs in $C_W(x)$ if and only if x occurs in $E_W(y)$)
(ii) $W \in \boldsymbol{SUB}[U]$
(iii) W is the least of all c-e structures R satisfying $V \leq R \leq U$ for all $V \in A$

Therefore $W = \sum A$. □

Before several properties of the least upper bounds of sets of firing components are stated, let us establish some general facts on the least upper bounds of sets of c-e structures:

Proposition 6.2 *(a) For any sets* $A, B \subseteq \mathbf{CE}$*:* $\sum(A \cup B) = \sum A + \sum B$ *provided that all the least upper bounds involved exist.*
(b) For any sets $A, B \subseteq \mathbf{CE}$*:* $A \subseteq B \Rightarrow \sum A \leq \sum B$ *provided that* $\sum A$ *and* $\sum B$ *exist.*
(c) $\forall j \in I : (U_j \leq V_j \Rightarrow \sum_{j \in I} U_j \leq \sum_{j \in I} V_j$ *if* $\sum_{j \in I} U_j$ *and* $\sum_{j \in I} V_j$ *exist.*
(d) $\sum \mathbf{CE}$ *does not exist.*

Proof **Of (a)** Since $\sum A \leq \sum(A \cup B)$ and $\sum B \leq \sum(A \cup B)$ then by (a) in Proposition 2.2: $\sum A + \sum B \leq \sum(A \cup B)$. $U \in A \cup B \Leftrightarrow U \in A \vee U \in B \Rightarrow$ (by (a) in Proposition 6.1) $U \in \boldsymbol{SUB}\left[\sum A\right] \vee U \in \boldsymbol{SUB}\left[\sum B\right] \Leftrightarrow U \in \boldsymbol{SUB}\left[\sum A\right] \cup \boldsymbol{SUB}\left[\sum B\right] \Rightarrow U \in \boldsymbol{SUB}\left[\sum A + \sum B\right] \Leftrightarrow U \leq \sum A + \sum B$. By definition of the least upper bound, $\sum(A \cup B)$ is the least of all V's satisfying $U \leq V$, thus $\sum(A \cup B) \leq \sum A + \sum B$. Therefore $\sum(A \cup B) = \sum A + \sum B$.

Of (b) $A \subseteq B \Rightarrow B = A \cup B \Rightarrow \sum B = \sum(A \cup B) =$ (by (a)) $\sum A + \sum B$. Thus, $\sum A \leq \sum B$.

Of (c) $\forall j \in I : U_j \leq V_j \Leftrightarrow \forall j \in I : (V_j = U_j + V_j) \Rightarrow \sum_{j \in I} V_j = \sum_{j \in I} (U_j + V_j)$. From the general law for least upper bounds: $\sum_{i \in I} \sum_{j \in J} W_{ij} = \sum_{j \in J} \sum_{i \in I} W_{ij}$ $(W_{ij} \in \mathbf{CE})$, it follows that $\sum_{j \in I} (U_j + V_j) = \sum_{j \in I} U_j + \sum_{j \in I} V_j$. Thus $\sum_{j \in I} V_j = \sum_{j \in I} U_j + \sum_{j \in I} V_j$, which is equivalent to $\sum_{j \in I} U_j \leq \sum_{j \in I} V_j$.

Of (d) Let $\Omega = (C_\Omega, E_\Omega) = \sum \mathbf{CE}[\mathbb{X}]$ exist. Obviously $car(\Omega) = \mathbb{X}$. Since $\Omega \in \mathbf{CE}[\mathbb{X}]$, then for each node $x \in car(\Omega)$, the polynomial $E_\Omega(x)$ is finite, thus it has a finite number of arguments. Consider a node $a \in car(\Omega)$. Obviously, there exist $b \in car(\Omega)$, which does not occur in $E_\Omega(a)$ and consequently a does not occur in $C_\Omega(b)$. Define $\Omega' = (C_{\Omega'}, E_{\Omega'})$ with:

$$C_{\Omega'}(x) = \begin{cases} C_\Omega(x) & \text{for } x \neq b \\ C_\Omega(x) + a & \text{for } x = b \end{cases} \qquad E_{\Omega'}(x) = \begin{cases} E_\Omega(x) & \text{for } x \neq a \\ E_\Omega(x) + b & \text{for } x = a \end{cases}$$

Certainly $\Omega' \in \mathbf{CE}[\mathbb{X}]$ (i.e. x occurs in $C_{\Omega'}(y)$ iff y occurs in $E_{\Omega'}(x)$) and $\Omega < \Omega'$—a contradiction with the assumption $\Omega = \sum \mathbf{CE}[\mathbb{X}]$. □

Some properties of the least upper bounds of (sets of) firing components are in:

Proposition 6.3 *(a) For each* $A \subseteq \boldsymbol{FC}$, *if* $\sum A$ *exists then* $A \subseteq \boldsymbol{FC}[\sum A]$

(b) For each $U \in \boldsymbol{CE}$ *there exists* $\sum \boldsymbol{FC}[U]$ *and* $\sum \boldsymbol{FC}[U] \leq U$. *Sometimes* $\sum \boldsymbol{FC}[U] \neq U$

(c) $\bigcup_{j\in I} \boldsymbol{FC}[U_j] \subseteq \boldsymbol{FC}[\sum_{j\in I} U_j]$ *(generalization of (m) in Proposition 2.2)*

(d) If $\sum \boldsymbol{FC}[U] = U$ *then* $\boldsymbol{FC}[U] \subseteq \boldsymbol{FC}[V] \Rightarrow U \leq V$

Proof **Of (a)** By assumptions and by (a) in Proposition 6.1: $A \subseteq \boldsymbol{SUB}[\sum A]$, $A \subseteq \boldsymbol{SUB}[\sum A] \cap \boldsymbol{FC} = \boldsymbol{FC}[\sum A]$.

Of (b) The existence of $\sum \boldsymbol{FC}[U]$ follows from (d) in Proposition 6.1. $\boldsymbol{SUB}[U] \cap \boldsymbol{FC} \subseteq \boldsymbol{SUB}[U] \Rightarrow$ (by (b) in Proposition 6.2) $\sum(\mathbf{SUB}[U] \cap \boldsymbol{FC}) \leq \sum \boldsymbol{SUB}[U]$. Thus, by (b) in Proposition 6.1: $\sum \boldsymbol{FC}[U] \leq U$. Let $U = \{a_{b\bullet c+b}, b^a_c, c^{a\bullet b}\}$. Then $\boldsymbol{FC}[U] = \{\{a_b, b^a\}\}$, $\sum \boldsymbol{FC}[U] = \{a_b, b^a\} \neq U$.

Of (c) By (c) in Proposition 6.1: $\boldsymbol{FC} \cap \bigcup_{j\in I} \boldsymbol{SUB}[U_j] \subseteq \boldsymbol{FC} \cap \boldsymbol{SUB}[\sum_{j\in I} U_j]$. Since $\boldsymbol{FC} \cap \bigcup_{j\in I} \boldsymbol{SUB}[U_j] = \bigcup_{j\in I}(\boldsymbol{FC} \cap \boldsymbol{SUB}[U_j])$, we get $\bigcup_{j\in I} \boldsymbol{FC}[U_j] \subseteq \boldsymbol{FC}[\sum_{j\in I} U_j]$.

Of (d) $\boldsymbol{FC}[U] \subseteq \boldsymbol{FC}[V] \Rightarrow$ (by (b) in Proposition 6.2) $\sum \boldsymbol{FC}[U] \leq \sum \boldsymbol{FC}[V]$. Hence, by assumption and by (b) we get $U \leq V$. □

Let us examine some lattice properties of c-e structures.

A partial order $(Z, \leq)$ is a *lattice* iff each two $x, y \in Z$ have the greatest lower bound (the greatest u with $u \leq x$, $u \leq y$) and the least upper bound (the least v with $x \leq v$, $y \leq v$) denoted respectively $glb(x, y)$ and $lub(x, y)$. The lattice is *distributive* if and only if $glb(x, lub(y, z)) = lub(glb(x, y), glb(x, z))$.

Theorem 6.1 $\langle \boldsymbol{CE}, \leq \rangle$*—the set of all c-e structures partially ordered by* $U \leq V$ *iff* $V = U + V$, *is a non-distributive lattice with the least element* θ *and with no greatest element.*

Proof For $U, V \in \boldsymbol{CE}$ let $W_1 = \sum(\boldsymbol{SUB}[U] \cap \boldsymbol{SUB}[V])$, $W_2 = \sum(\boldsymbol{SUB}[U] \cup \boldsymbol{SUB}[V])$. We show that $glb(U, V) = W_1$ and $lub(U, V) = W_2$. Let $A = \boldsymbol{SUB}[\mathrm{U}] \cap \boldsymbol{SUB}[V]$. Since $A \subseteq \boldsymbol{SUB}[U]$ and $A \subseteq \boldsymbol{SUB}[V]$, then, by (d) in Proposition 6.1, $\sum A$ exists in $\boldsymbol{SUB}[U]$ and in $\boldsymbol{SUB}[V]$, hence $\sum A$ exists in $\boldsymbol{SUB}[U] \cap$

$\boldsymbol{SUB}[V]$. Thus, $\sum A \leq U \wedge \sum A \leq V$ and for each $T \in A : \sum A \leq T \Rightarrow \sum A = T$ (because $\sum A$, as the least upper bound of A, is the maximal element in A). Therefore $(T \leq U \wedge T \leq V) \Rightarrow T \leq \sum A$, which means that $W_1 = \sum A = glb(U, V)$. $W_2 = \sum (\boldsymbol{SUB}[U] \cup \boldsymbol{SUB}[V]) =$ (by (a) in Proposition 6.2) $\sum \boldsymbol{SUB}[U] + \sum \boldsymbol{SUB}[V] =$ (by (b) in Proposition 6.1) $U + V = lub(U, V)$. Thus, $(\mathbf{CE}, \leq)$ is a lattice. It is not distributive: let $U = \{a_x, x^a\}$, $V = \{a_{x+y}, b_{x\bullet y}, x^{a\bullet b}, y^{a\bullet b}\}$, $W = \{a_{x\bullet y}, x^a, y^a\}$, thus $lub(V, W) = V + W = \{a_{x+y+x\bullet y}, b_{x\bullet y}, x^{a\bullet b+a}, y^{a\bullet b+a}\}$, $\boldsymbol{SUB}[U] \cap \boldsymbol{SUB}[V + W] = \{\theta, U\}$ hence $glb(U, V + W) = U$, $\boldsymbol{SUB}[U] \cap \boldsymbol{SUB}[V] = \{\theta\}$, hence $glb(U, V) = \theta$, $\boldsymbol{SUB}[U] \cap \boldsymbol{SUB}[W] = \{\theta\}$, hence $glb(U, W) = \theta$. Therefore, $U = glb(U, lub(V, W)) \neq lub(glb(U, V), glb(U, W)) = \theta$. Obviously θ, is the least element in CE. By (d) in Proposition 6.2, there is no greatest element in **CE**. This ends the proof. □

Existence of the greatest lower bound and the least upper bound of every pair U, V of c-e structures, as well as equalities:
$glb(U, V) = \sum (\boldsymbol{SUB}[U] \cap \boldsymbol{SUB}[V])$
$lub(V, V) = \sum (\boldsymbol{SUB}[U] \cup \boldsymbol{SUB}[V])$
is ascertained in the proof of Theorem 6.1.

6.2 Properties of the Greatest Lower Bound of Sets of c-e Structures

For a set of c-e structures $A \subseteq \boldsymbol{CE}$, the greatest lower bound of A, denoted by $\prod A$, is the greatest (wrt $\leq$) c-e structure V with $V \leq U$ for each $U \in A$. In addition, let $\prod \emptyset = \theta$. Obviously, $\prod A$ always exists and $glb(U, V) = \prod\{U, V\}$.

Although $U + V = lub(U, V)$, this is not the case that $U \bullet V = glb(U, V)$. Denote the glb operation in the infix form, and let us examine the lattice properties of c-e structures more thoroughly.

Definition 6.1 $U \circ V = \sum (\boldsymbol{SUB}[U] \cap \boldsymbol{SUB}[V]) = glb(U, V)$. □

Evidently $U \circ V \in \boldsymbol{CE}$, so in the resulting structure, x occurs in $C_{U \circ V}(y)$ iff y occurs in $E_{U \circ V}(x)$. Thus, the glb operation on c-e structures yields correct c-e structure.

Notice that whereas operation "•" composes (or synchronises—see Figs. 1.3 and 4.11) c-e structures, the operation "∘" removes from them some parts.

The Theorem 6.1 may be now reformulated: $\langle \boldsymbol{CE}, +, \circ \rangle$ is a non-distributive lattice with the least element θ and without the greatest.

Proposition 6.4 *For all* $U, V \in \boldsymbol{CE}$*:*

(a) $U \circ (V + W) \geq U \circ V + U \circ W$ *(compare (c) in Proposition 2.2)*
(b) If $\boldsymbol{SUB}[V] \cup \boldsymbol{SUB}[W] = \boldsymbol{SUB}[V + W]$ *then* $U \circ (V + W) = U \circ V + U \circ W$
(c) $U \leq V \Leftrightarrow U = U \circ V$

Proof **Of (a)** $U \circ (V + W) = \sum (\boldsymbol{SUB}[U] \cap \boldsymbol{SUB}[V + W]) \geq$ (by (i) in Proposition 2.2 and (b) in Proposition 6.2) $\sum (\boldsymbol{SUB}[U] \cap (\boldsymbol{SUB}[V] \cup \boldsymbol{SUB}[W])) = \sum[(\boldsymbol{SUB}[U] \cap \boldsymbol{SUB}[V]) \cup (\boldsymbol{SUB}[U] \cap \boldsymbol{SUB}[W])] =$ (by (a) in Proposition 6.2) $\sum(\boldsymbol{SUB}[U] \cap \boldsymbol{SUB}[V]) + \sum(\boldsymbol{SUB}[U] \cap \boldsymbol{SUB}[W]) = U \circ V + U \circ W$.

Of (b) Like that of (a), but replace "≥" with "="—by assumption.

Of (c) $U \leq V \Leftrightarrow \boldsymbol{SUB}[U] \subseteq \boldsymbol{SUB}[V]$ (by (h) in Proposition 2.2) $\Leftrightarrow$
$\boldsymbol{SUB}[U] \cap \boldsymbol{SUB}[V] = \boldsymbol{SUB}[U] \Leftrightarrow$
$\sum(\boldsymbol{SUB}[U] \cap \boldsymbol{SUB}[V]) = \sum \boldsymbol{SUB}[U] \Leftrightarrow$
$U \circ V = U$ (by (b) in Proposition 6.1). □

It is worth noting that although not always $\boldsymbol{FC}[U] \cup \boldsymbol{FC}[V] = \boldsymbol{FC}[U + V]$ holds (see (n) in Proposition 2.2), the *glb* enjoys property stated in Proposition 6.5.

Proposition 6.5 *For all* $U, V \in \boldsymbol{CE}$: $\boldsymbol{FC}[U] \cap \boldsymbol{FC}[V] = \boldsymbol{FC}[U \circ V]$

Proof Let $Q \in \boldsymbol{FC}[U] \cap \boldsymbol{FC}[V]$. Then $Q \in \boldsymbol{SUB}[U] \cap \boldsymbol{SUB}[V]$, which, by definition of the least upper bound yields $Q \leq \sum(\boldsymbol{SUB}[U] \cap \boldsymbol{SUB}[V]) = U \circ V$. Hence, $Q \in \boldsymbol{FC}[U \circ V]$, since $\boldsymbol{FC}[U \circ V] = \boldsymbol{SUB}[U \circ V] \cap \boldsymbol{FC}$. Therefore $\boldsymbol{FC}[U] \cap \boldsymbol{FC}[V] \subseteq \boldsymbol{FC}[U \circ V]$. Inclusion $\boldsymbol{FC}[U \circ V] \subseteq \boldsymbol{FC}[U] \cap \boldsymbol{FC}[V]$ follows from $U \circ V \leq U$, $U \circ V \leq V$, by (l) in Proposition 2.2. This ends the proof. □

Theorem 6.2 *For each* $U \in \boldsymbol{CE}$, $U \neq \theta$, $\langle \boldsymbol{SUB}[U], +, \circ \rangle$ *is a non-distributive, complete lattice with the least and greatest elements* θ *and* U *respectively (a lattice is complete if any of its non-empty subsets has the greatest lower bound and the least upper bound).*

Proof We show completeness only (remaining properties follow directly from Theorem 6.1 and definition of $\boldsymbol{SUB}[U]$). Let $A \subseteq \boldsymbol{SUB}[U]$ and $B = \{W \mid \forall V \in A : W \leq V\}$. Obviously, $B \subseteq \boldsymbol{SUB}[U]$ thus, by (d) in Proposition 6.1, there exists $\sum B$. Since $W \in B$ implies $W \leq V$ for each $V \leq A$, then by (c) in Proposition 6.2: $\sum B \leq V$ for each $V \in A$. But $\sum B$ is the greatest of all W's satisfying $\forall V \in A : W \leq V$, thus $\sum B$ is the greatest lower bound of the set A. The least upper bound of A, i.e. $\sum A$ exists by (d) in Proposition 6.1. Therefore, $\prod A \in \boldsymbol{SUB}[U]$ and $\sum A \in \boldsymbol{SUB}[U]$. □

Now, we are in a position to state some properties of the quasi semiring of c-e structures over $\mathbb{X}$ apart from equations in Proposition 2.1, additionally, by equations in Proposition 6.6.

Proposition 6.6 *For all* $U, V, W \in \boldsymbol{CE}[\mathbb{X}]$

($\circ$) $\theta \circ U = U \circ \theta = \theta$

($\circ\circ$) $U \circ U = U$

($\circ\circ\circ$) $U \circ V = V \circ U$

($\circ\circ\circ\circ$) $U \circ (V \circ W) = (U \circ V) \circ W$

($+\circ$) *If* $\boldsymbol{SUB}[V] \cup \boldsymbol{SUB}[W] = \boldsymbol{SUB}[V + W]$ *then* $U \circ (V + W) = U \circ V + U \circ W$

Proof Equations ($\circ$), ($\circ\circ$), ($\circ \circ \circ$) are evident, implication ($+\circ$)—proved in Proposition 6.4(b), so, let us prove associativity ($\circ \circ \circ\circ$).

To this end notice that

$$\boldsymbol{SUB}[\sum(\boldsymbol{SUB}[V] \cap \boldsymbol{SUB}[W])] = \boldsymbol{SUB}[V] \cap \boldsymbol{SUB}[W].$$

Therefore, $U \circ (V \circ W) = \sum(\ \boldsymbol{SUB}[U] \cap\ \boldsymbol{SUB}[\sum(\ \boldsymbol{SUB}[V] \cap\ \boldsymbol{SUB}[W])]) = \sum(\ \boldsymbol{SUB}[U] \cap\ \boldsymbol{SUB}[V] \cap\ \boldsymbol{SUB}[W]) = \sum(\ \boldsymbol{SUB}[\sum(\ \boldsymbol{SUB}[U] \cap\ \boldsymbol{SUB}[V])] \cap \boldsymbol{SUB}[W]) = (U \circ V) \circ W.$ □

Chapter 7
Further Structural Properties—Decomposition

When the concept of cause-effect structures originated in late 1980s and 1990s, it was believed that each c-e structure could be represented as a combination of single arrows $a \longrightarrow b$ by means of operations "+" and "$\bullet$". In other words, that each c-e structure could be decomposed into single arrows, i.e. that the algebra of c-e structures is generated by single arrows. Abraham [1] of the Ben Gurion University was the first to find a 7-nodes non-decomposable c-e structure, then Marek Raczunas of the Warsaw University found a 6-nodes non-decomposable c-e structure [7]. Furthermore, it turned out that each c-e structure $U = \langle C_U, E_U \rangle$ can be transformed into the "two-level" c-e structure $U' = \langle C_{U'}, E_{U'} \rangle$ with $C_{U'}(x) = \theta \Leftrightarrow E_{U'}(x) \neq \theta$ for each $x \in car(U')$, retaining the property of decomposability/non-decomposability (though not behaviour). Moreover, there are infinitely many of non-decomposable, called "*prime c-e structures*" in the quasi-semiring, but any two-level c-e structure with 2, 3, 4, 5 nodes is decomposable. This is the subject of this chapter.

Definition 7.1 A c-e structure U is *decomposable into a sum or product* if there are c-e structures $V, W \in \boldsymbol{CE}[\mathbb{X}]$ with $\theta \neq V \neq U \neq W \neq \theta$ and $U = V + W$ or $U = V \bullet W$. U is *decomposable* if it is decomposable into a sum or a product. U is *non-decomposable,* called ***prime***, if it is not decomposable. U is *decomposable into single arrows* iff U may be represented by an expression with arrows $x \rightarrow y$ (where $x, y \in \mathbb{X}$) as operands and + and/or $\bullet$ as operators. □

Decomposition of c-e structures is not, in general, unique, for example $U = \{a^{\theta}_{x+y}, b^{\theta}_{x \bullet y}, x^{a \bullet b}_{\theta}, y^{a \bullet b}_{\theta}\}$ may be decomposed into single arrows either as $U = ((a \rightarrow x) + (a \rightarrow y)) \bullet (b \rightarrow x) \bullet (b \rightarrow y)$ or as $U = ((a \rightarrow x) \bullet (b \rightarrow x) + (a \rightarrow y)) \bullet (b \rightarrow y)$.

Due to the Theorem 7.2, we focus on the bipartite c-e structures $U = (C, E)$, with $C_U(x) = \theta \Leftrightarrow E_U(x) \neq \theta$ for each $x \in car(U)$. It is convenient to present such "*two-level*" c-e structures in a "matrix notation", as shows the example:

$$\{a^{\theta}_{x+y}, b^{\theta}_{x \bullet y}, x^{a \bullet b}_{\theta}, y^{a \bullet b}_{\theta}\} = \begin{pmatrix} a_{x+y} & b_{x \bullet y} \\ x^{a \bullet b} & y^{a \bullet b} \end{pmatrix} \quad \begin{matrix} \text{level 1} \\ \text{level 2} \end{matrix}$$

L. Czaja, *Cause-Effect Structures*, Lecture Notes in Networks and Systems 45, https://doi.org/10.1007/978-3-030-20461-7_7

Note that each firing component is a two-level c-e structure.

Theorem 7.1 *There exist infinitely many prime c-e structures. Moreover, for each $m \geq 6$ there exists a prime two-level c-e structure U with $|car(U)| = m$.*

Proof Consider the infinite sequence U_3, U_4, ..., U_n, ... where:

$$U_n = \begin{pmatrix} a_{\mathbf{1+2\bullet 3\bullet ...\bullet n}} & b_{\mathbf{1\bullet 2\bullet 3\bullet ...\bullet n}} & c_{\mathbf{1\bullet 2\bullet 3\bullet ...\bullet n}} & & \\ \mathbf{1}^{a\bullet b+c} & \mathbf{2}^{a\bullet c+b} & \mathbf{3}^{b\bullet c+a} & \mathbf{4}^{b\bullet c+a} & \ldots\ldots\ \mathbf{n}^{b\bullet c+a} \end{pmatrix}$$

for $n \geq 3$. U_n is a two-level c-e structure: the upper level consists of nodes a, b, c, the lower level—of nodes **1,2,3,.....,n** (bold numerals as names of nodes). Let $U_n = (C, E)$. Then $C(a) = C(b) = C(c) = E(\mathbf{j}) = \theta$ for $\mathbf{j} = \mathbf{1,2,3,.....,n}$ and $C(\mathbf{j}) = b \bullet c + a$ for $\mathbf{j} = \mathbf{3, 4, ..., n}$. We show that U_n is prime for all $n \geq 3$.

Suppose $U_n = V \bullet W$ with $\theta \neq V \neq U_n$ and $\theta \neq W \neq U_n$. Neither $E(a)$ nor $C(\mathbf{j})$ for $\mathbf{j} = \mathbf{1,2,3,.....,n}$ can be represented as a product of polynomials distinct from θ, thus each of the nodes a,**1,2,3,.....,n** belongs either to $car(V)$ or to $car(W)$ but no one of them may belong to both. But all **1,2,3,.....,n** must jointly with node a belong either to $car(V)$ or to $car(W)$ since a occurs in all $C(\mathbf{j})$ for $\mathbf{j} = \mathbf{1,2,3,.....,n}$. If, for instance, a,**1,2,3,.....,n** $\in car(V)$ then $b, c \in car(V)$ since **1,2,3,.....,n** occur in $E(b)$ and in $E(c)$. But $b, c \notin car(W)$ because, otherwise at least one of **1,2,3,.....,n** would belong to $car(W)$. Therefore $car(W) = \emptyset$ and $car(V) = car(U_n)$, implying $W = \theta$ and $V = U_n$—the contradiction. Hence, U_n is non-decomposable into a product.

Suppose $U_n = V + W$ with $\theta \neq V \neq U_n$ and $\theta \neq W \neq U_n$. First, we show the following implications:
$car(V) \neq \emptyset \Rightarrow car(V) = car(U_n)$ and $car(W) \neq \emptyset \Rightarrow car(W) = car(U_n)$ (⊛)
Indeed, if $a \in car(V)$ then either $E_V(a) = \mathbf{1}$ or $E_V(a) = \mathbf{2 \bullet 3 \bullet ... \bullet n}$ or $E_V(a) = \mathbf{1 + 2 \bullet 3 \bullet ... \bullet n}$. Each of these cases implies that any other node of U_n is in $car(V)$. Similar reasoning with other nodes yields the same conclusion and the same applies to W. Second, we show that $E_V(\xi) = E(\xi)$ and $C_V(\xi) = C(\xi)$ for each $\xi \in car(U_n)$. This would mean that $V = U_n$—the contradiction. Consider cases:

Case $\xi = a$
$a \in car(V) \Rightarrow$(by implications (⊛)) $c \in car(V) \Rightarrow E_V(c) = \mathbf{1 \bullet 2 \bullet 3 \bullet ... \bullet n} \Rightarrow$ c occurs in $C_V(\mathbf{2}) \Rightarrow a$ occurs in $C_V(\mathbf{2}) \Leftrightarrow \mathbf{2}$ occurs in $E_V(a) \Rightarrow \mathbf{2, 3, ..., n}$ occur in $E_V(a)$. On the other hand, $a \in car(V) \Rightarrow$(by (⊛)) $b \in car(V) \Rightarrow E_V(b) = \mathbf{1 \bullet 2 \bullet 3 \bullet ... \bullet n} \Rightarrow b$ occurs in $C_V(\mathbf{1}) \Rightarrow a$ occurs in $C_V(\mathbf{1}) \Leftrightarrow \mathbf{1}$ occurs in $E_V(a)$. Concluding: $E_V(a) = \mathbf{1 + 2 \bullet 3 \bullet ... \bullet n} = E(a)$.

Cases $\xi = b$ and $\xi = c$ are obvious.

Case $\xi = \mathbf{1}$
$\mathbf{1} \in car(V) \Rightarrow$(by (⊛)) $c \in car(V) \Rightarrow E_V(c) = \mathbf{1 \bullet 2 \bullet 3 \bullet ... \bullet n} \Rightarrow c$ occurs in $C_V(\mathbf{1})$. On the other hand, $\mathbf{1} \in car(V) \Rightarrow$(by (⊛)) $b \in car(V) \Rightarrow E_V(b) = \mathbf{1 \bullet 2 \bullet 3 \bullet ... \bullet n} \Rightarrow b$ occurs in $C_V(\mathbf{1}) \Rightarrow a, b$ occur in $C_V(\mathbf{1})$. Concluding: $C_V(\mathbf{1}) = a \bullet b + c = C(\mathbf{1})$.

Case $\xi = \mathbf{2}$
$\mathbf{2} \in car(V) \Rightarrow$(by (⊛)) $b \in car(V) \Rightarrow E_V(b) = \mathbf{1} \bullet \mathbf{2} \bullet \mathbf{3} \bullet \ldots \bullet \mathbf{n} \Rightarrow b$ occurs in $C_V(\mathbf{2})$. On the other hand, $\mathbf{2} \in car(V) \Rightarrow$(by (⊛)) $c \in car(V) \Rightarrow E_V(c) = \mathbf{1} \bullet \mathbf{2} \bullet \mathbf{3} \bullet \ldots \bullet \mathbf{n} \Rightarrow c$ occurs in $C_V(\mathbf{2}) \Rightarrow a, c$ occur in $C_V(\mathbf{2})$.Concluding: $C_V(\mathbf{2}) = a \bullet c + b = C(\mathbf{2})$.

Case $\xi = \mathbf{3}$
$\mathbf{3} \in car(V) \Rightarrow$(by (⊛)) $a \in car(V) \Rightarrow$(by **Case** $\xi = a$) $E_V(a) = \mathbf{1} + \mathbf{2} \bullet \mathbf{3} \bullet \ldots \bullet \mathbf{n} \Rightarrow a$ occurs in $C_V(\mathbf{3})$. On the other hand, $\mathbf{3} \in car(V) \Rightarrow$(by (⊛)) $b \in car(V) \Rightarrow E_V(b) = \mathbf{1} \bullet \mathbf{2} \bullet \mathbf{3} \bullet \ldots \bullet \mathbf{n} \Rightarrow b$ occurs in $C_V(\mathbf{3}) \Rightarrow b, c$ occur in $C_V(\mathbf{3})$. Concluding: $C_V(\mathbf{3}) = b \bullet c + a = C(\mathbf{3})$.

Cases $\xi = \mathbf{4}, \mathbf{5}, \ldots, \mathbf{n}$ are treated identically to **Case** $\xi = \mathbf{3}$.

All this means that $C_V = C$ and $E_V = E$, thus $V = U_n$—the contradiction. Hence, U_n is non-decomposable into a sum. The same reasoning applies to W, which ends the proof. □

Obviously, each arrow $x \to y$ is a prime c-e structure, thus we have:

Corollary *Any c-e structure is either prime or it is a combination (by means of* + *and* •*) of prime c-e structures.*

Theorem 7.2 states that examination of composability of c-e structures may be reduced to examination of composability of some two-level c-e structures.

Theorem 7.2 *For any c-e structure U there exists a two-level c-e structure $U' = \langle C_{U'}, E_{U'} \rangle$ (thus, $C_{U'}(x) = \theta \Leftrightarrow E_{U'}(x) \neq \theta$ for each $x \in car(U')$), such that U is prime if and only if U' is prime.*

Proof Suppose the universe $\mathbb{X}$ of node names contains neither underlined nor over-barred symbols. Assign to each $x \in \mathbb{X}$ two symbols $\underline{x}$ and $\overline{x}$ and let $\underline{\mathbb{X}} = \{\underline{x}: x \in \mathbb{X}\}$, $\overline{\mathbb{X}} = \{\overline{x}: x \in \mathbb{X}\}$. Given a c-e structure $U \in \boldsymbol{CE}[\mathbb{X}]$, construct a c-e structure $U' \in \boldsymbol{CE}[\underline{\mathbb{X}} \cup \overline{\mathbb{X}}]$ as follows:
(a) Replace each $x \in car(U)$ with $\underline{x}$ and $\overline{x}$; they will make $car(U')$.
(b) Let $C_{U'}(\underline{x}) = E_{U'}(\overline{x}) = \theta$
(c) Let $C_{U'}(\overline{x})$ be obtained from $C_U(x)$ by replacement of all y occuring in $C_U(x)$ with $\underline{y}$
(d) Let $E_{U'}(\underline{x})$ be obtained from $E_U(x)$ by replacement of all y occuring in $E_U(x)$ with $\overline{y}$
Note that U' is a two level c-e structure indeed, that is $C_{U'}(\xi) = \theta \Leftrightarrow E_{U'}(\xi) \neq \theta$ for each $\xi \in car(U')$ and ξ occurs in $C_{U'}(\eta)$ iff η occurs in $E_{U'}(\xi)$. These replacements define a function f: $\boldsymbol{CE}[\mathbb{X}] \to \boldsymbol{CE}[\underline{\mathbb{X}} \cup \overline{\mathbb{X}}]$, such that $f(U) = U'$. We show that f is a one-to-one function such that $f(V + W) = f(V) + f(W)$ and $f(V \bullet W) = f(V) \bullet f(W)$ as well as $f^{-1}(V' + W') = f^{-1}(V') + f^{-1}(W')$ and $f^{-1}(V' \bullet W') = f^{-1}(V') \bullet f^{-1}(W')$, where V' and W' are obtained from V and W likewise U' from U above. Let $U_1 = (C_{U_1}, E_{U_1}) \neq U_2 = (C_{U_2}, E_{U_2})$. Then, $\exists x \in \mathbb{X} : (C_{U_1}(x) \neq C_{U_2}(x) \vee E_{U_1}(x) \neq E_{U_2}(x)) \Leftrightarrow \exists x \in \mathbb{X} : (C_{f(U_1)}(\overline{x}) \neq C_{f(U_2)}(\overline{x}) \vee$

$E_{f(U_1)}(\underline{x}) \neq E_{f(U_2)}(\underline{x})) \Leftrightarrow (C_{f(U_1)}, E_{f(U_1)}) \neq (C_{f(U_2)}, E_{f(U_2)}) \Leftrightarrow f(U_1) \neq f(U_2)$. Thus, f is one-to-one. Let $V = (C_V, E_V)$ and $W = (C_W, E_W)$. For demonstration of $f(V + W) = f(V) + f(W)$ it is to be shown:

(C+) $C_{f(V+W)}(\xi) = C_{f(V)}(\xi) + C_{f(W)}(\xi)$

(E+) $E_{f(V+W)}(\xi) = E_{f(V)}(\xi) + E_{f(W)}(\xi)$

for each $\xi \in \underline{\mathbb{X}} \cup \overline{\mathbb{X}}$ and the same for "•".

Let $\xi = \overline{x}$ for a certain $x \in \mathbb{X}$. Then, by (b): $E_{f(V+W)}(\xi) = E_{f(V)}(\xi) = E_{f(W)}(\xi) = \theta$, thus (E+) holds. By definition of f, $C_{f(V+W)}(\overline{x})$, is the polynomial $C_{V+W}(x) = C_V(x) + C_W(x)$ in which every argument y has been replaced with $\underline{y}$. But replacement of every y in $C_V(x)$ and in $C_W(x)$ with $\underline{y}$ yields $C_{f(V)}(\overline{x})$ and $C_{f(W)}(\overline{x})$ respectively, thus, (C+) holds. Let $\xi = \underline{x}$ for a certain $x \in \mathbb{X}$. Repeat the above reasoning interchanging E with C and $\overline{x}$ with $\underline{x}$. Apply the whole reasoning to "•" analogously. Equations $f^{-1}(V' + W') = f^{-1}(V') + f^{-1}(W')$ and $f^{-1}(V' \bullet W') = f^{-1}(V') \bullet f^{-1}(W')$ are demonstrated similarly, using the results for f. Therefore, we have shown that U is decomposable into a sum or product if and only if $f(U) = U'$ is decomposable, which ends the proof. □

As an example, consider the non-two-level, 3-node, decomposable c-e structure U in Fig. 7.1 and its transformation into the two-level c-e structure $U' = f(U)$ in Fig. 7.2.

$U = (a \longrightarrow b) \bullet (a \longrightarrow c) + (a \longrightarrow c) \bullet (b \longrightarrow c) + (c \longrightarrow a) + (c \longrightarrow b)$

U' is decomposable into arrows as follows:

$U' = (\underline{b} \rightarrow \overline{c}) \bullet ((\underline{a} \rightarrow \overline{c}) + (\underline{a} \rightarrow \overline{c}) \bullet (\underline{a} \rightarrow \overline{b})) + (\underline{c} \rightarrow \overline{b}) + (\underline{c} \rightarrow \overline{a})$.

An example of a non-two-level, 3-node and prime c-e structure U, is depicted in Fig. 7.3, and its transformation into the two-level c-e structure $U'' = f(U)$—in Fig. 7.4.

This c-e structure is prime, since its transformation described in the proof of Theorem 7.2 yields the two-level c-e structure depicted in Fig. 7.4.

The c-e structure in Fig 7.4 is prime because it is the renamed c-e structure U_3 from the proof of Theorem 7.1.

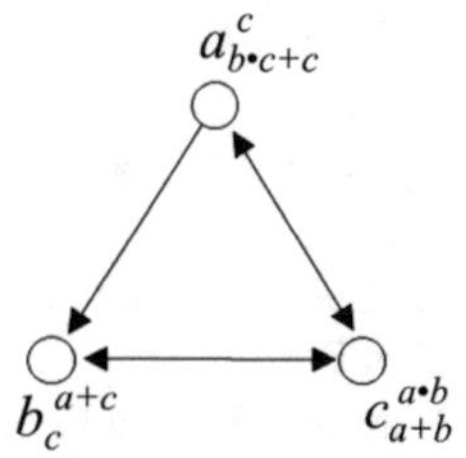

Fig. 7.1 Non-two-level, 3-node, decomposable c-e structure

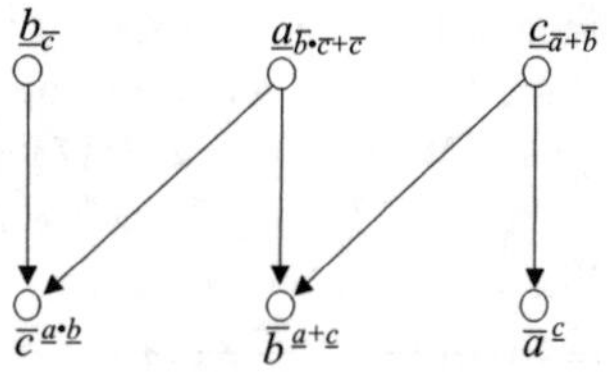

Fig. 7.2 Two-level c-e structure $U' = f(U)$ constructed in the proof of Theorem 7.2

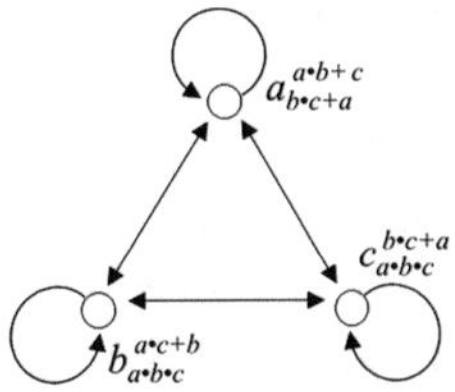

Fig. 7.3 Non-two-level, 3-node, prime c-e structure U

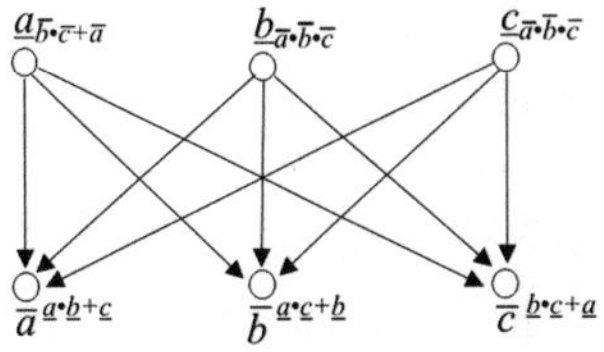

Fig. 7.4 Two-level c-e structure $U'' = f(U)$ constructed in the proof of Theorem 7.2

Before coming to a surprising fact stated in Theorem 7.3, let us notice a simpler property in Proposition 7.1.

Proposition 7.1 *Any two-level c-e structure with a single node at one level is decomposable into arrows.*

Proof Such c-e structure is of the form $U = \{a_{K(x1,x2,...,xn)}, x1^a, x2^a, ..., xn^a\}$, where $K(x1, x2, ..., xn)$ is a polynomial of arguments $x1, x2, ..., xn$. It may be represented as $U = \overline{K}(a \to x1,\ a \to x2, ..., a \to xn)$ where $\overline{K}$ is K with each $x1, x2, ..., xn$ replaced with $a \to x1,\ a \to x2, ..., a \to xn$ respectively, and +, • meant as operations on c-e structures. □

Theorem 7.3 *Any two level c-e structure with 3, 4 or 5 nodes is decomposable into arrows.*

Proof For a two-level c-e structure of the type (1, n) or (n, 1), that is, with a single node at one level and n nodes at the other, the theorem follows from Proposition 7.1. A proof for type (2, 2) procedes as follows. Let $U = \begin{pmatrix} a_I & b_J \\ x^K & y^L \end{pmatrix}$, where I, J are arbitrary polynomials of x, y and K, L are arbitrary polynomials of a, b. We reduce the number of cases to be considered to a few only, by identifying some rules. Quite general and evident rules pertain to symmetry, for instance, having proved decomposition of U for particular I and J, we abandon cases where they are interchanged. The more specific rules are: U is decomposable provided that at least one of the following conditions (Rules 1–4) holds. By $< M >$ is denoted the set of all arguments occurring in the polynomial M. One has to take care that at every stage of decomposition, correct c-e structures $U = (C, E)$ are obtained, that is such that x occures in $C(y)$ iff y occurs in $E(x)$.

Rule 1: A polynomial $\Delta \in \{I, J, K, L\}$ is a sum $\Delta = G + H$ with $G \neq \Delta \neq H$ and $< G >=< H >$
Rule 2: $| < \Delta > | \leq 1$ for a certain $\Delta \in \{I, J, K, L\}$
Rule 3: $I = J = x \bullet y$
Rule 4: $I = J = x + y$

Justification of the Rule1: Let $I = G + H$. Then $U = \begin{pmatrix} a_G & b_J \\ x^K & y^L \end{pmatrix} + \begin{pmatrix} a_H & b_J \\ x^K & y^L \end{pmatrix}$ and both summands are correct c-e structures—by assumption $< G >=< H >$ and are distinct from U, since $G \neq I \neq H$.

Justification of the Rule 2: Let $< I >= \{x\}$. Then $U = \begin{pmatrix} a_x & b_J \\ x^K & y^b \end{pmatrix}$. Due to the Rule 1, there is no need to consider $K = a \bullet b + a + b$; case $K = b$ is impossible, $K = a$—trivial. For every $K \in \{a + b,\ a \bullet b,\ a \bullet b + a,\ a \bullet b + b\}$ U is decomposable:

$$\begin{pmatrix} a_x & b_J \\ x^{a+b} & y^b \end{pmatrix} = \begin{pmatrix} a_x \\ x^a \end{pmatrix} + \begin{pmatrix} & b_J \\ x^b & y^b \end{pmatrix}$$

$$\begin{pmatrix} a_x & b_J \\ x^{a\bullet b} & y^b \end{pmatrix} = \begin{pmatrix} a_x \\ x^a \end{pmatrix} \bullet \begin{pmatrix} & b_J \\ x^b & y^b \end{pmatrix}$$

$$\begin{pmatrix} a_x & b_J \\ x^{a\bullet b+a} & y^b \end{pmatrix} = \begin{pmatrix} a_x \\ x^a \end{pmatrix} + \begin{pmatrix} a_x & b_J \\ x^{a\bullet b} & y^b \end{pmatrix}$$

$$\begin{pmatrix} a_x & b_J \\ x^{a\bullet b+b} & y^b \end{pmatrix} = \begin{pmatrix} a_x & b_J \\ x^{a\bullet b} & y^b \end{pmatrix} + \begin{pmatrix} & b_J \\ x^b & y^b \end{pmatrix}$$

Justification of the Rule 3:

$$U = \begin{pmatrix} a_{x\bullet y} & b_{x\bullet y} \\ x^K & y^L \end{pmatrix} = \begin{pmatrix} a_x & b_x \\ x^K & \end{pmatrix} \bullet \begin{pmatrix} a_y & b_y \\ & y^L \end{pmatrix}$$

Justification of the Rule 4:

$$U = \begin{pmatrix} a_{x+y} & b_{x+y} \\ x^K & y^L \end{pmatrix} = \begin{pmatrix} a_x & b_x \\ x^K & \end{pmatrix} + \begin{pmatrix} a_y & b_y \\ & y^L \end{pmatrix}$$

Cases of decomposition not encompassed by Rules 1–4 are pointed to by entries of the following Table:

$E(b)$ \ $E(a)$	$x + y$	$x \bullet y$	$x \bullet y + x$	$x \bullet y + y$
$x + y$	$R4$	$SC1$	$SC1$	$SC1$
$x \bullet y$	$C1$	$R3$	$SC2$	$SC2$
$x \bullet y + x$	$C1$	$C2$	$C3$	$SC3$
$x \bullet y + y$	$C1$	$C2$	$C3$	$C4$

Entries point to Rules and Cases of decomposition:
Ri—Rule i ($i = 3, 4$)
Cj—Case j ($j = 1, 2, 3, 4$) listed in the tables that follow
SCj—Case symmetric to Cj, that is with x and y interchanged

Case 1: $E(a) = x + y$, $E(b)$—specified in the first column,
$K, L \in \{a + b,\ a \bullet b,\ a \bullet b + a,\ a \bullet b + b\}$,
Addends and factors in the column U—decomposable by means of the appropriate Rules above.

$E(b)$	$C(x)$	$C(y)$	U
$x \bullet y$	$a + b$	L	$\begin{pmatrix} a_x \\ x^a \end{pmatrix} + \begin{pmatrix} a_y & b_{x\bullet y} \\ x^b & y^L \end{pmatrix}$
	$a \bullet b$	L	$\begin{pmatrix} b_x \\ x^b \end{pmatrix} \bullet \begin{pmatrix} a_{x+y} & b_y \\ x^a & y^L \end{pmatrix}$
	$a \bullet b + a$	L	$\begin{pmatrix} a_x \\ x^a \end{pmatrix} + \begin{pmatrix} a_{x+y} & b_{x\bullet y} \\ x^{a\bullet b} & y^L \end{pmatrix}$
	$a \bullet b + b$	L	$\begin{pmatrix} b_x \\ x^b \end{pmatrix} \bullet \begin{pmatrix} a_{x+y} & b_{x\bullet y} \\ x^{a+b} & y^L \end{pmatrix}$
$x \bullet y + x$	K	L	$\begin{pmatrix} a_x\ b_x \\ x^K \end{pmatrix} + \begin{pmatrix} a_{x+y} & b_{x\bullet y} \\ x^K & y^L \end{pmatrix}$
$x \bullet y + y$	K	L	$\begin{pmatrix} a_y\ b_y \\ y^L \end{pmatrix} + \begin{pmatrix} a_{x+y} & b_{x\bullet y} \\ x^K & y^L \end{pmatrix}$

Case 2: $E(a) = x \bullet y$, $E(b)$—specified in the first column,
$L \in \{a + b,\ a \bullet b,\ a \bullet b + a,\ a \bullet b + b\}$
Addends and factors in the column U—decomposable by means of the appropriate Rules above.

$E(b)$	$C(x)$	$C(y)$	U
	$a + b$	L	$\begin{pmatrix} b_x \\ x^b \end{pmatrix} + \begin{pmatrix} a_{x\bullet y} & b_{x\bullet y} \\ x^{a+b} & y^L \end{pmatrix}$
	$a \bullet b$	L	$\begin{pmatrix} a_x \\ x^a \end{pmatrix} \bullet \begin{pmatrix} a_y & b_{x\bullet y+x} \\ x^b & y^L \end{pmatrix}$
$x \bullet y + x$	$a \bullet b + a$	L	$\begin{pmatrix} a_x \\ x^a \end{pmatrix} \bullet \begin{pmatrix} a_{x\bullet y} & b_{x\bullet y+x} \\ x^{a+b} & y^L \end{pmatrix}$
	$a \bullet b + b$	L	$\begin{pmatrix} b_x \\ x^b \end{pmatrix} + \begin{pmatrix} a_{x\bullet y} & b_{x\bullet y} \\ x^{a\bullet b} & y^L \end{pmatrix}$
$x \bullet y + y$	analogously: interchange x and y		

Case 3: $E(a) = x \bullet y + x$, $E(b)$—specified in the first column,
$K, L \in \{a + b,\ a \bullet b,\ a \bullet b + a,\ a \bullet b + b\}$
Addends and factors in the column U—decomposable by means of the appropriate Rules above.

$E(b)$	$C(x)$	$C(y)$	U
$x \bullet y + x$	K	L	$\begin{pmatrix} a_x\ b_x \\ x^K \end{pmatrix} + \begin{pmatrix} a_{x\bullet y} & b_{x\bullet y} \\ x^K & y^L \end{pmatrix}$
$x \bullet y + y$	K	L	$\begin{pmatrix} a_{x\bullet y+x} & b_{x\bullet y} \\ x^K & y^L \end{pmatrix} + \begin{pmatrix} a_{x\bullet y} & b_{x\bullet y+y} \\ x^K & y^L \end{pmatrix}$

Case 4: $E(a) = x \bullet y + x$—analogous to Case 3.
The proof of decomposability of two-level c-e structures of type (2, 3) or (3, 2), required consideration of several hundred cases. It is in [2, 4] and is not quoted here. □

7.1 Bilogic c-e Structures

An important class is created by the *bilogic* c-e structures (for the bilogic graphs cf. [3, 5, 6]). Any such c-e structure is decomposable as stated in Theorem 7.4.

Definition 7.2 A c-s structure $U = \langle C, E \rangle$ is *bilogic* if none of polynomials $C(x)$, $E(x)$ comprises jointly both operators "+" and "•". □

For example, Figs. 2.7, 4.5, 5.1 depict bilogic c-e structures.

Theorem 7.4 *Any bilogic c-e structure is decomposable into single arrows (or is already such or is neutral* θ*).*

Proof It suffices to consider two-level c-e structures, because any bilogic one, remains bilogic, after transformation (described in the proof of Theorem 7.2) into two-level c-e structure. Let

$$U = \begin{pmatrix} a_{E(a)} & b_{E(b)} & \ldots\ldots & c_{E(c)} \\ x^{C(x)} & y^{C(y)} & \ldots\ldots & z^{C(z)} \end{pmatrix}$$

be bilogic. The exhaustive cases to be considered are the following:

Case 1: Some polynomials $E(a), E(b), \ldots, E(c)$ from level 1 do not comprise "+", thus each of them is a product of nodes (or a single node). Let for instance $E(a) = x \bullet y \bullet \ldots \bullet z$. Consider sub-cases:
(1+) some $C(x), C(y), \ldots, C(z)$ do not comprise "+"
(1•) all the $C(x), C(y), \ldots, C(z)$ do not comprise "•"
In (1+) let for instance $C(x) = a \bullet K$ and polynomial K does not comprise "+". Then

$$U = \begin{pmatrix} a_x \\ x^a \end{pmatrix} \bullet \begin{pmatrix} a_{y\bullet\ldots\bullet z} & b_{E(b)} & \ldots\ldots & c_{E(c)} \\ x^K & y^{C(y)} & \ldots\ldots & z^{C(z)} \end{pmatrix}$$

In (1•) let for instance $C(x) = a + K$, $C(y) = a + L, \ldots\ldots, C(z) = a + M$ where $K, L, \ldots, M$ do not comprise "•". Then:

$$U = \begin{pmatrix} a_x \\ x^a \end{pmatrix} \bullet \begin{pmatrix} a_y \\ y^a \end{pmatrix} \bullet \ldots\ldots \bullet \begin{pmatrix} a_z \\ z^a \end{pmatrix} + \begin{pmatrix} a_{y\bullet\ldots\bullet z} & b_{E(b)} & \ldots\ldots & c_{E(c)} \\ x^K & y^L & \ldots\ldots & z^M \end{pmatrix}$$

The right factor of U in the sub-case (1+) and the addend in the sub-case (1•), as bilogic, may be decomposed again in the way described. For the following sub-cases (2+) and (2•), dual to (1+) and (1•)—reasoning analogous:

Case 2: interchange in Case 1: "+" with "•", product with sum, (1+) with (2•) and (1•) with (2+). □

7.2 Structural Deadlock Versus Decomposition

A c-e structure U is a *structural deadlock* if no firing component is in U, that is, when $\boldsymbol{FC}[U] = \emptyset$. In such c-e structures no change of state is possible. The decomposable c-e structure $\{a^{\theta}_{x+y}, b^{\theta}_{x\bullet y}, x^{a\bullet b}_{\theta}, y^{a\bullet b}_{\theta}\}$ as well as all prime c-e structures U_n $(n \geq 3)$ used in the proof of Theorem 7.1 are examples of structural deadlocks. Another example is the (decomposable) c-e structure $\{a^{\theta}_{b\bullet c}, b^{a}_{c}, c^{a\bullet b}_{\theta}\}$. Thus, structural deadlocks may be either prime or decomposable. The converse question: *is any prime c-e structure a structural deadlock?* is answered by the following example:

$$U = \begin{pmatrix} a_{x\bullet y+x\bullet z+u} & b_{x\bullet y+z\bullet u} & c_{x\bullet u+y\bullet z} & \\ x^{a\bullet b+b\bullet c} & y^{a\bullet b+c} & z^{a\bullet b\bullet c} & u^{a\bullet b\bullet c} \end{pmatrix}.$$

U **is prime**. Indeed, let $U = V \bullet W$ with $\emptyset \neq V \neq U \neq W \neq \theta$. Neither $E_U(b)$ nor $C_U(y)$ may be represented as a product of polynomials distinct from θ, thus either $b, y \in car(V)$ or $b, y \in car(W)$, but $b \notin car(V) \cap car(W)$ and $y \notin car(V) \cap car(W)$. If, for instance, $b, y \in car(V)$ then $a, c \in car(V)$ since y occurs in $E_U(a)$ and in $E_U(c)$. But, $a, c, x, z, u \notin car(W)$ since otherwise $y \in car(W)$ or $b \in car(W)$ would hold. Therefore, $car(W) = \emptyset$ and $car(V) = car(U)$ implying $W = \theta$ and $V = U$—the contradiction. Hence U is not decomposable into a product. Let $U = V + W$ with $\emptyset \neq V \neq U \neq W \neq \theta$. If at least one of polynomials $E_V(a)$, $E_V(b)$, $E_V(c)$ contains z or u (or both) or at least one of polynomials $C_V(x)$, $C_V(y)$, $C_V(z)$, $C_V(u)$, contains c then $car(V) = car(U)$. But this implies $V = U$, because every monomial occurring in V must occur in U too (since monomials cannot be split between V and W). Suppose that $E_V(a)$, $E_V(b)$ contain neither z nor u, that is $E_V(a) = E_V(b) = x \bullet y$. This implies $C_V(x) = C_V(y) = a \bullet b$. Similarly, if $C_V(x) = C_V(y) = a \bullet b$ then $E_V(a) = E_V(b) = x \bullet y$. Therefore $V = \{a^{\theta}_{x\bullet y}, b^{\theta}_{x\bullet y}, x^{a\bullet b}_{\theta}, y^{a\bullet b}_{\theta}\}$. But then $W = U$, which contradicts $\emptyset \neq W \neq U$. Concluding, U is prime.

U **is not a structural deadlock**. Indeed, $U = \{a^{\theta}_{x\bullet y}, b^{\theta}_{x\bullet y}, x^{a\bullet b}_{\theta}, y^{a\bullet b}_{\theta}\} + U$, thus $\boldsymbol{FC}[U] = \{\{a^{\theta}_{x\bullet y}, b^{\theta}_{x\bullet y}, x^{a\bullet b}_{\theta}, y^{a\bullet b}_{\theta}\}\} \neq \emptyset$. Therefore, we have the following Theorem 7.5.

Theorem 7.5 *The notions of structural deadlock and decomposition of c-e structures are independent.* □

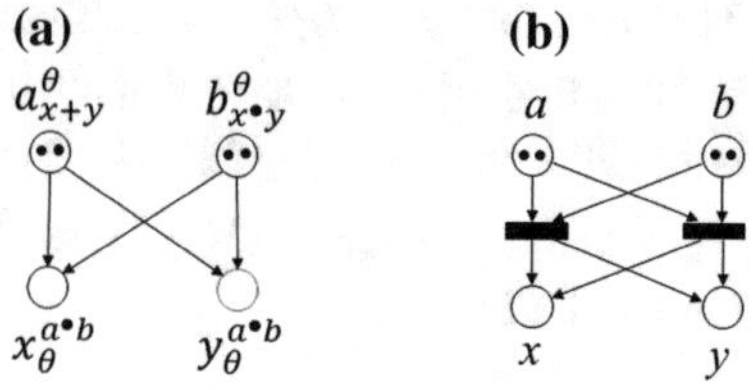

Fig. 7.5 **a** Structural deadlock; **b** Petri net (seemingly equivalent to (**a**), but in fact—not)—each transition can fire

Remark Remember that firing component is a structural (syntactic) notion, independent of state (Definition 2.5). Its definition, as such, was possible due to inadmissible the so-called non-simple c-e structures—the counterparts of non-simple Petri nets, where more than one transition has the same preset and postset. For example, the multivalued (Definition 3.3) c-e structure in Fig. 7.5a has no firing components (is a structural deadlock), whereas the non-simple Petri net in Fig. 7.5b has two transitions. Thus, the the c-e structure cannot change state $\overline{s}(a) = \overline{s}(b) = 2$, $\overline{s}(x) = \overline{s}(y) = 0$, but the Petri net can (assuming sufficient capacity of nodes-places). The behavioural deadlock in c-e structures will be dealt with in Chap. 8.

References

1. Abraham U (1991) On decomposition of cause-effect structures. A seminar at Department of Mathematics and Physics, University of Gdansk
2. Czaja L (1995) Decomposition of cause-effect structures. Institute of Informatics, Warsaw University TR 95–07(207)
3. Cerf VG (1972) Multiprocessors, semaphores, and a graph model of computation. PhD thesis, Computer Science Department, University of California, Los Angeles
4. Deminet J (1991) Structural properties of cause-effect nets. PhD thesis (in Polish), Warsaw University
5. Gostelow KP (1971) Flow of control, resource allocation, and the proper termination of programs. PhD thesis, Computer Science Department, University of California, Los Angeles
6. Peterson JL, Bredt TH (1974) A comparison of models of parallel computation. In: Proceedings of the IFIP Congress 74, vol 3 (Mathematical aspects of information processing), Stockholm, Sweden
7. Raczunas M (1995) Processes and c-e structures. Preparatory typescript for PhD thesis, Warsaw University

Chapter 8
Semantic Properties of c-e Structures

A few of the semantic properties collected in this chapter are general properties of relations. Nonetheless, they are quoted as useful means for analysis of the c-e structures' behaviour. Some properties, specific to c-e structures and mentioned in Propositions 2.3, 3.1 (Chaps. 2, 3), are important for comparison with facts established in other models of concurrency, especially in Petri nets. So they are repeated and proved. All semantics: $[[U]]$, $[[U]]^*$, $[[U]]_{par}$ $[[U]]^*_{par}$ of a c-e structure U are defined in Chaps. 2 and 3.

Proposition 8.1 *For any c-e structures* U, $V \in \boldsymbol{CE}$:
(a) $U \le V \Rightarrow \boldsymbol{FC}[U] \subseteq \boldsymbol{FC}[V] \Rightarrow [[U]] \subseteq [[V]] \Rightarrow [[U]]^* \subseteq [[V]]^*$
(b) $\boldsymbol{FC}[U] \cup \boldsymbol{FC}[V] = \boldsymbol{FC}[U+V] \Rightarrow [[U]] \cup [[V]] = [[U+V]]$ *but not conversely.*
(c) $\boldsymbol{FC}[U] \cup \boldsymbol{FC}[V] = \boldsymbol{FC}[U+V]$ *and* $[[U]]^* \cup [[V]]^* = [[U+V]]^*$ *are unrelated by implication.*
(d) $\boldsymbol{FC}[U] \cup \boldsymbol{FC}[V] = \boldsymbol{FC}[U+V]$ *and* $[[U]]_{par} \cup [[V]]_{par} = [[U+V]]_{par}$ *are unrelated by implication.*
(e) $\boldsymbol{FC}[U] \cup \boldsymbol{FC}[V] = \boldsymbol{FC}[U+V]$ *and* $[[U]]^*_{par} \cup [[V]]^*_{par} = [[U+V]]^*_{par}$ *are unrelated by implication.*

Proof **Of (a)** The first "⇒"—evident, the second "⇒": by definition of semantics we have $[[U]] = \bigcup_{Q \in \boldsymbol{FC}[U]} [[Q]]$, $[[V]] = \bigcup_{Q \in \boldsymbol{FC}[V]} [[Q]]$ and because $\boldsymbol{FC}[U] \subseteq \boldsymbol{FC}[V]$ implies $\boldsymbol{FC}[V] = \boldsymbol{FC}[U] \cup \boldsymbol{FC}[V]$, thus $[[V]] = \bigcup_{Q \in \boldsymbol{FC}[U] \cup \boldsymbol{FC}[V]} [[Q]] = \bigcup_{Q \in \boldsymbol{FC}[U]} [[Q]] \cup \bigcup_{Q \in \boldsymbol{FC}[V]} [[Q]] = [[U]] \cup [[V]]$. Therefore $[[U]] \subseteq [[V]]$. The third "⇒" is the general property of relations.
Of (b) $[[U]] \cup [[V]] =$ (by definition of semantics) $\bigcup_{Q \in \boldsymbol{FC}[U]} [[Q]] \cup \bigcup_{Q \in \boldsymbol{FC}[V]} [[Q]] = \bigcup_{Q \in \boldsymbol{FC}[U] \cup \boldsymbol{FC}[V]} [[Q]] =$ (by assumption) $\bigcup_{Q \in \boldsymbol{FC}[U+V]} [[Q]] = [[U+V]]$. A counter example to the converse implication is: $U = \{a_{x\bullet y+x}, b_y, x^a, y^{a\bullet b}\}$, $V = \{a_x, b_{x\bullet y+y},$

L. Czaja, *Cause-Effect Structures*, Lecture Notes in Networks and Systems 45, https://doi.org/10.1007/978-3-030-20461-7_8

$x^{a\bullet b}, y^b\}$. Then: $U + V = \{a_{x\bullet y+x}, b_{x\bullet y+y}, x^{a\bullet b+a}, y^{a\bullet b+b}\}$, $\boldsymbol{FC}[U] = \{\{a_{x\bullet y}, b_y, x^a, y^{a\bullet b}\}, \{a_x, x^a\}\}$, $\boldsymbol{FC}[V] = \{\{a_x, b_{x\bullet y}, x^{a\bullet b}, y^b\}, \{b_y, y^b\}\}$, $\boldsymbol{FC}[U + V] = \{\{a_{x\bullet y}, b_{x\bullet y}, x^{a\bullet b}, y^{a\bullet b}\}\} \cup \boldsymbol{FC}[U] \cup \boldsymbol{FC}[V]$. $[[U]] = [[\{a_{x\bullet y}, b_{x\bullet y}, x^{a\bullet b}, y^{a\bullet b}\}]] \cup [[U]] \cup [[V]]$. But $[[\{a_{x\bullet y+x}, b_y, x^a, y^{a\bullet b}\}]] = [[\{a_x, b_{x\bullet y}, x^{a\bullet b}, y^b\}]] = [[\{a_{x\bullet y}, b_{x\bullet y}, x^{a\bullet b}, y^{a\bullet b}\}]]$. Therefore, $[[U]] \cup [[V]] = [[U + V]]$ and $\boldsymbol{FC}[U] \cup \boldsymbol{FC}[V] \neq \boldsymbol{FC}[U + V]$.

Of (c) A counter example to $\boldsymbol{FC}[U] \cup \boldsymbol{FC}[V] = \boldsymbol{FC}[U + V] \Rightarrow [[U]]^* \cup [[V]]^* = [[U + V]]^*$ is: $U = \{x_y, y^x\}$, $V = \{y_z, z^y\}$. Indeed, $U + V = \{x_y, y_z^x, z^y\}$, $\boldsymbol{FC}[U] = \{\{x_y, y^x\}\}$, $\boldsymbol{FC}[V] = \{\{y_z, z^y\}\}$, $\boldsymbol{FC}[U + V] = \{\{x_y, y^x\}, \{y_z, z^y\}\}$. Thus, $\boldsymbol{FC}[U] \cup \boldsymbol{FC}[V] = \boldsymbol{FC}[U + V]$ and $[[U]]^* \cup [[V]]^* \neq [[U + V]]^*$, since $(\{x\}, \{z\}) \notin [[U]]^* \cup [[V]]^*$ and $(\{x\}, \{z\}) \in [[U + V]]^*$.

A counter example to $[[U]]^* \cup [[V]]^* = [[U + V]]^* \Rightarrow \boldsymbol{FC}[U] \cup \boldsymbol{FC}[V] = \boldsymbol{FC}[U + V]$ is: $U = \{a_{x\bullet y+y}, x^{a+y}, y_x^a\}$, $V = \{a_x, b_{x\bullet y}, x^{a\bullet b}, y^b\}$, $U + V = \{a_{x\bullet y+x+y}, b_{x\bullet y}, x^{a\bullet b+a+y}, y_x^{a+b}\}$, $\boldsymbol{FC}[U] = \{\{a_{x\bullet y}, x^a, y^a\}, \{a_y, y^a\}, \{y_x, x^y\}\}$, $\boldsymbol{FC}[V] = \{V\}$ $[[U]]^* = \{(\{a, b, x\}, \{a, x, y\}), (\{a, b, y\}, \{a, b, x\}), (\{a, b, y\}, \{b, x, y\}), (\{a, b\}, \{b, x, y\}), (\{a, b\}, \{b, y\}), (\{b, y\}, \{b, x\}), (\{a, b\}, \{b, x\}), (\{a, x\}, \{x, y\}), (\{a, y\}, \{a, x\})$ $(\{a, y\}, \{x, y\}), (\{a\}, \{x, y\}), (\{a\}, \{y\}), (\{y\}, \{x\}), (\{a\}, \{x\})\}$. $[[V]]^* = \{(\{a, b\}, \{x, y\})\}$, $[[U]]^* \cup [[V]]^* = [[U + V]]^*$, $\boldsymbol{FC}[U] \cup \boldsymbol{FC}[V] \neq \boldsymbol{FC}[U + V]$. For simplicity, we assumed $\mathbb{X} = car(U) \cup car(V)$.

Of (d) The counter example to the implication $\boldsymbol{FC}[U] \cup \boldsymbol{FC}[V] = \boldsymbol{FC}[U + V] \Rightarrow [[U]]_{par} \cup [[V]]_{par} = [[U + V]]_{par}$ is: $U = \{a_x, x^a\}$, $V = \{b_y, y^b\}$, since $U + V = \{a_x, x^a, b_y, y^b\}$ and, as may be easily checked, $\boldsymbol{FC}[U] \cup \boldsymbol{FC}[V] = \boldsymbol{FC}[U + V]$ but $[[U]]_{par} \cup [[V]]_{par} \neq [[U + V]]_{par}$. The counter example to the implication $[[U]]_{par} \cup [[V]]_{par} = [[U + V]]_{par} \Rightarrow \boldsymbol{FC}[U] \cup \boldsymbol{FC}[V] = \boldsymbol{FC}[U + V]$ is: $U = \{a_x, b_{x\bullet y+x+y}, x^{a+b}, y^b\}$, $V = \{a_x, b_x, x^{a\bullet b}\}$, since $U + V = \{a_x, b_{x\bullet y+x+y}, x^{a\bullet b+a+b}, y^b\}$ and $[[U]]_{par} \cup [[V]]_{par} = [[U + V]]_{par}$ (note that $(\{a, b\}, \{x, y\}) \in [[U]]_{par}$) *but* $\boldsymbol{FC}[U] \cup \boldsymbol{FC}[V] \neq \boldsymbol{FC}[U + V]$.

Of (e) Follows from (c) and from equality $[[U]]^* = [[U]]_{par}^*$ proved in (b) in Proposition 8.2. □

Proposition 8.2 *(a)* $[[U]] \subseteq [[U]]_{par}$ *but the reverse inclusion not always holds*
(b) $[[U]]^* = [[U]]_{par}^*$
(c) $[[U]] \cup [[V]] \subseteq [[U + V]]$, *but the reverse inclusion not always holds*
(d) $[[U]] \cup [[V]] \subseteq [[U]]^* \cup [[V]]^* \subseteq ([[U]] \cup [[V]])^* \subseteq [[U + V]]^*$ *but none of the inclusions "⊆" may be replaced with equality "=".*

Proof **Of (a)** If $Q \in \boldsymbol{FC}[U]$ transforms a state s into t in the semantics $[[U]]$ then $\{Q\}$ does it in the semantics $[[U]]_{par}$. A counter example to the reverse inclusion: $U = \{a_x, x^a, b_y, y^b\}$.

Of (b) Inclusion $[[U]]^* \subseteq [[U]]_{par}^*$ follows from (a). It remains to prove inclusion $[[U]]_{par}^* \subseteq [[U]]^*$. Let $(s, t) \in [[U]]_{par}^*$. If $s = t$ then $(s, t) \in [[U]]^*$. If $s \neq t$ then there are states $s_0, s_1, ..., s_n$ $(n > 0)$ with $s = s_0$, $t = s_n$, $(s_j, s_{j+1}) \in [[U]]_{par}$, $j = 0, 1, ..., n - 1$. Thus, there are $G_0, G_1, ..., G_{n-1} \subseteq \boldsymbol{FC}[U]$, every G_j being pairwise detached (pwd—Definition 2.8), with $(s_j, s_{j+1}) \in [[G_j]]_{par}$. Let $G_j = \{Q_{j1}, Q_{j2}, ..., Q_{jp_j}\}$, where $Q_{ji} \in \boldsymbol{FC}[U]$, for $i = 1, 2, ..., p_j$ $(p_j > 0)$.

Since ${}^{\bullet}Q^{\bullet}_{ji} \cap {}^{\bullet}Q^{\bullet}_{jk} = \emptyset$ for $i \neq k$ (because G_j is pwd), there are states $s_{j0}, s_{j1}, \ldots, s_{jp_j}$ with $s_j = s_{j0}$, $s_{j+1} = s_{jp_j}$, $(s_{ji}, s_{ji+1}) \in [[Q_{ji+1}]]$, for $i = 0, 1, \ldots, p_j - 1$. Hence $(s_{ji}, s_{ji+1}) \in [[U]]$, thus $(s_j, s_{j+1}) \in [[U]]^*$ and since this holds for every $j = 0, 1, \ldots, n-1$, we get $(s, t) \in [[U]]^*$.
Of (c) $U \leq U + V$ and $V \leq U + V$ thus, by (a) in Proposition 8.1: $[[U]] \subseteq [[U + V]]$ and $[[V]] \subseteq [[U + V]]$, hence (c) holds.
Of (d) Follows from general properties of relations and (c). □

Notice that implication $\boldsymbol{FC}[U] \cup \boldsymbol{FC}[V] = \boldsymbol{FC}[U + V] \Rightarrow [[U]] \cup [[V]] = [[U + V]]$ (point (b) in Proposition 8.1) states what might be called a ***conditional compositionality*** of semantics [[ ]], with respect to "∪" (on the semantic level) and "+" (on the structural level). It states that behaviour in effect of acting of one firing component of a sum (+) of c-e structures, is obtained by making the union (∪) of behaviour of its summands, provided that the summation (+) does not create new firing components. This does not concern semantics $[[\ ]]^*$ nor $[[\ ]]_{par}$. Notice that Propositions 8.1 and 8.2 have been proved without getting into various definitions of firing component's semantics, i.e. various definitions of relation $[[Q]]$, see the Remark in Chap. 3.

8.1 A Note on Safety

The properties stated in Propositions 8.1, 8.2 (and generalizations in the further Proposition 8.3) may be used for verification of the so called safety property of c-e structures. Unlike in the Petri nets theory but in consent with the theory of programming, by *safety* is meant here the property: each state reachable from a given initial state must satisfy a certain condition $\Gamma_{(U,s_0)}$, which specifies the intended behaviour of the c-e structure U with initial state s_0. That is, one has to verify $\forall s : [(s_0, s) \in [[U]]^* \Rightarrow \Gamma_{(U,s_0)}(s)]$, or the similar statement for semantics $[[U]]_{par}$. Decomposition of U into substructures and decomposition of $\Gamma_{(U,s_0)}$ into relevant subconditions, then application of the mentioned Propositions, offers a method of safety analysis. Note that $\Gamma_{(U,s_0)}$ is an *invariant* of the U's activity, that is, it should hold in any state s reachable from s_0 in U. Another property of parallel systems is *liveness,* considered in the further part of this chapter. To illustrate the safety analysis, consider the following example.

Example (Verification of Readers/Writers c-e Structure Depicted in Chap. 3, Fig. 3.6)

This c-e structure contains twelve firing components depicted in Fig. 8.1.

Summing up all the firing components we get the whole Readers/Writers c-e structure in Fig. 3.6:

$$RW = \sum_{j=1}^{3} RW[j], \text{ where } RW[j] = \mathrm{W}{\downarrow}(j) + \mathrm{W}{\uparrow}(j) + \mathrm{R}{\downarrow}(j) + \mathrm{R}{\uparrow}(j)$$

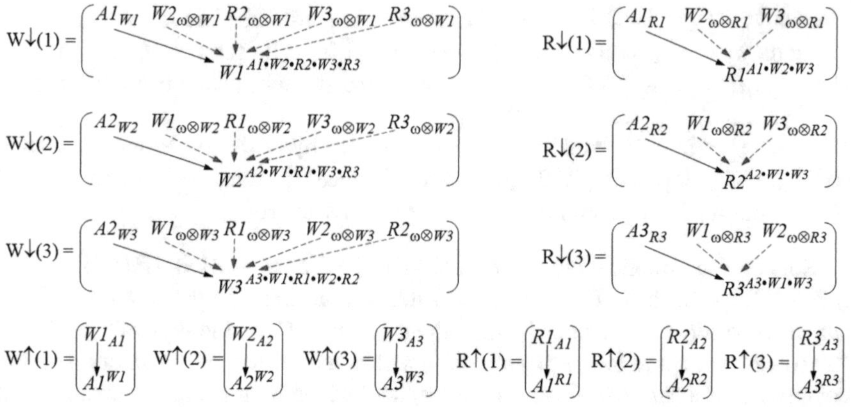

$$W{\downarrow}(1) = \begin{pmatrix} A1_{W1} & W2_{\omega\otimes W1} & R2_{\omega\otimes W1} & W3_{\omega\otimes W1} & R3_{\omega\otimes W1} \\ & & W1^{A1\cdot W2\cdot R2\cdot W3\cdot R3} \end{pmatrix} \qquad R{\downarrow}(1) = \begin{pmatrix} A1_{R1} & W2_{\omega\otimes R1} & W3_{\omega\otimes R1} \\ & R1^{A1\cdot W2\cdot W3} \end{pmatrix}$$

$$W{\downarrow}(2) = \begin{pmatrix} A2_{W2} & W1_{\omega\otimes W2} & R1_{\omega\otimes W2} & W3_{\omega\otimes W2} & R3_{\omega\otimes W2} \\ & & W2^{A2\cdot W1\cdot R1\cdot W3\cdot R3} \end{pmatrix} \qquad R{\downarrow}(2) = \begin{pmatrix} A2_{R2} & W1_{\omega\otimes R2} & W3_{\omega\otimes R2} \\ & R2^{A2\cdot W1\cdot W3} \end{pmatrix}$$

$$W{\downarrow}(3) = \begin{pmatrix} A2_{W3} & W1_{\omega\otimes W3} & R1_{\omega\otimes W3} & W2_{\omega\otimes W3} & R2_{\omega\otimes W3} \\ & & W3^{A3\cdot W1\cdot R1\cdot W2\cdot R2} \end{pmatrix} \qquad R{\downarrow}(3) = \begin{pmatrix} A3_{R3} & W1_{\omega\otimes R3} & W2_{\omega\otimes R3} \\ & R3^{A3\cdot W1\cdot W3} \end{pmatrix}$$

$$W{\uparrow}(1) = \begin{pmatrix} W1_{A1} \\ A1^{W1} \end{pmatrix} \quad W{\uparrow}(2) = \begin{pmatrix} W2_{A2} \\ A2^{W2} \end{pmatrix} \quad W{\uparrow}(3) = \begin{pmatrix} W3_{A3} \\ A3^{W3} \end{pmatrix} \quad R{\uparrow}(1) = \begin{pmatrix} R1_{A1} \\ A1^{R1} \end{pmatrix} \quad R{\uparrow}(2) = \begin{pmatrix} R2_{A2} \\ A2^{R2} \end{pmatrix} \quad R{\uparrow}(3) = \begin{pmatrix} R3_{A3} \\ A3^{R3} \end{pmatrix}$$

Fig. 8.1 The set $FC[RW]$ of all firing components of the Readers/Writers c-e structure. Firing components $W{\downarrow}(j)$ and $R{\downarrow}(j)$ denote, respectively, the begin, whereas $W{\uparrow}(j)$ and $R{\uparrow}(j)$ the end of writing and reading, by agent of number $j = 1, 2, 3$. The nodes with subscripts $\omega\otimes W(j)$ and $\omega\otimes R(j)$ are inhibitors

Thus, the summation has not created new firing components, besides those in Fig. 8.1.

So, we have $FC[RW] = \bigcup_{j=1}^{3} RW[j]$ and by virtue of point (b) in Proposition 8.1 (and

its generalization in Proposition 8.3), $[[RW]] = \left[\left[\bigcup_{j=1}^{3} RW[j]\right]\right]$. The initial state s_0

is: $s_0(Aj) = 1$, $s_0(Wj) = s_0(Rj) = 0$, for $j = 1, 2, 3$. According to specification, if an agent of number $j = 1, 2, 3$ is writing, the remaining agents are precluded from writing and reading, but reading may proceed in parallel. Thus, the condition for exclusive writing by agent j is: $\neg\exists k, l : (l \neq j \wedge (s(Rk) = 1 \vee s(Wl) = 1))$ for each state s, reachable form s_0.

It is to be proved validity of the formula:

$$\forall s : \big[(s_0, s) \in [[RW]]^* \Rightarrow$$

$$(\forall j : s(Wj) = 1 \Rightarrow \neg\exists k, l : (l \neq j \wedge (s(Rk) = 1 \vee s(Wl) = 1)))\big] \qquad \textbf{SPEC}$$

which specifies safe behaviour of the c-e structure in Fig. 3.6.

$[[RW]]^*$ is the reflexive and transitive closure of $[[RW]]$, that is, $(s_0, s) \in [[RW]]^*$ iff $s_0 = s$ or there is a sequence of states $s_0, s_1, ..., s_{n-1}, s_n$ with $s = s_n$ and $(s_i, s_{i+1}) \in [[RW]]$ for $i = 0, 1, ..., n - 1$. So, it is to be proved that if this sequence, for $j = 1, 2, 3$, satisfies $s_n(Wj) = 1$ then, for every $k, l = 1, 2, 3$ with $l \neq j$, the equalities $s_n(Wl) = 0$ and $s_n(Rk) = 0$ are fulfilled. Induction:

(a) If $n = 1$ then the only firing component transforming s_0 into s_1 is either $W{\downarrow}(j)$ or $R{\downarrow}(j)$ for a certain j, thus, by Definition 3.4 of semantics: $(s_0, s_1) \in [[W{\downarrow}(j)]]$ or $(s_0, s_1) \in [[R{\downarrow}(j)]]$. Thus, for $s = s_1$, the formula **SPEC** is fulfilled in both cases.

(b) Hypothesis: let for any prefix $s_0, s_1, ..., s_{i-1}, s_i$ $(1 \leq i < n)$ of the computation $s_0, s_1, ..., s_i, ..., s_{n-1}, s_n$, the following hold:
$(s_0, s_1) \in [[RW]]$, $(s_1, s_2) \in [[RW]]$, ..., $(s_{i-1}, s_i) \in [[RW]]$, which means $(s_0, s_i) \in [[RW]]^*$ and validity of **SPEC** for $s = s_i$, in particular for $s = s_{n-1}$. It is to be proved that this holds for $s = s_n$. Passing from s_{n-1} to s_n in the computation $s_0, s_1, ..., s_{n-1}, s_n$ is performed by one of firing components among twelve in Fig. 8.1, in accordance with Definition 3.4. If, for $j = 1, 2, 3,$ this firing component is:

- $W{\downarrow}(j)$ then $s_{n-1} = s_0$ (agent j starts writing)
- $R{\downarrow}(j)$ then $s_{n-1}(Wk) = s_n(Wk) = 0$ for $k = 1, 2, 3$ and $s_{n-1}(Aj) = s_n(Rj) = 1$, $s_{n-1}(Rj) = s_n(Aj) = 0$ (agent j starts reading)
- $W{\uparrow}(j)$ then $s_{n-1}(Wj) = 1$ and $s_n = s_0$ (agent j completes writing)
- $R{\uparrow}(j)$ then $s_{n-1}(Wk) = s_n(Wk) = 0$ for $k = 1, 2, 3$ and $s_{n-1}(Rj) = s_n(Aj) = 1$, $s_{n-1}(Aj) = s_n(Rj) = 0$ (agent j completes reading)

Therefore $(s_{n-1}, s_n) \in [[RW]]$, which implies validity of **SPEC** for $s = s_n$ in all these cases.

End of Verification.

8.2 Semantic Properties of Least Upper Bounds of Sets of c-e Structures

Proposition 8.3 *For any family of c-e structures* $\{U_j\}_{j \in I}$ *such that* $\sum_{j \in I} U_j$ *exists and any c-e structure* U*:*
(a) $\bigcup_{j \in I}[[U_j]] \subseteq [[\sum_{j \in I} U_j]]$ *(generalization of (c) in Proposition 8.2)*
(b) $\bigcup_{j \in I} \boldsymbol{FC}[U_j] = \boldsymbol{FC}[\sum_{j \in I} U_j] \Rightarrow \bigcup_{j \in I}[[U_j]] = [[\sum_{j \in I} U_j]]$ *(generalization of (b) in Proposition 8.1)*
(c) $[[\sum \boldsymbol{FC}[U]]] = [[U]]$
(d) Properties (a) and (c) but not (b) hold for semantics $[[\ ]]_{par}$

Proof **Of (a)** For any $k \in I$: $U_k \leq \sum_{j \in I} U_j \Rightarrow$ (by (a) in Proposition 8.1) $[[U_k]] \subseteq [[\sum_{j \in I} U_j]]$, hence (a).

Of (b) $\bigcup_{j \in I}[[U_j]] =$ (by definition of semantics) $\bigcup_{j \in I} \bigcup_{Q \in \boldsymbol{FC}[U_j]} [[Q]] = \bigcup_{Q \in \bigcup_{j \in I} \boldsymbol{FC}[U_j]} [[Q]]$
$=$ (by assumption) $\bigcup_{Q \in \boldsymbol{FC}[\sum_{j \in I} U_j]} [[Q]] =$ (by definition of semantics) $[[\sum_{j \in I} U_j]]$.

Of (c) By (b) in Proposition 5.3 there exists $\sum \boldsymbol{FC}[U]$ and $\sum \boldsymbol{FC}[U] \leq U$, thus, by (a) in Proposition 8.1: $[[\sum \boldsymbol{FC}[U]]] \subseteq [[U]]$. Reverse inclusion is obtained as follows: $[[U]] =$ (by definition of semantics) $\bigcup_{Q \in \boldsymbol{FC}[U]} [[Q]] \subseteq$ (by (a)) $[[\sum_{Q \in \boldsymbol{FC}[U]} Q]] = [[\sum \boldsymbol{FC}[U]]]$. Therefore (c) holds.

Of (d)—analogously to (a) and (c). □

A comment on (c): although, in general, $\sum \boldsymbol{FC}[U] \neq U$, these two c-e structures behave identically. This means that parts of U absent in $\sum \boldsymbol{FC}[U]$ do not affect behaviour of U, thus, from the behavioural point of view, may be disregarded. An example is $U_1 = \{a_{b \bullet c+b}, b^a_c, c^{a \bullet b}\}$ which may be reduced to $V_1 = \{a_b, b^a\}$, since U_1 and V_1 behave identically (note that $\{a_{b \bullet c}, b^a_c, c^{a \bullet b}\}$ is a structural deadlock). However, c-e structures $W = U_1 + \{a_c, b_c, c^{a+b}\}$ and $W' = V_1 + \{a_c, b_c, c^{a+b}\}$ behave differently: the firing component $\{a_{b \bullet c}, b^a, c^a\}$, present in W, is absent in W'. This shows that, in general $[[U_1]] = [[V_1]] \wedge [[U_2]] = [[V_2]]$ does not imply $[[U_1 + U_2]] = [[V_1 + V_2]]$. However, the conditional implication is stated in Proposition 8.4.

Proposition 8.4 *For any c-e structures* $U_1, U_2, V_1, V_2 \in \boldsymbol{CE}$ *satisfying* $\boldsymbol{FC}[U_1] \cup \boldsymbol{FC}[U_2] = \boldsymbol{FC}[U_1 + U_2]$ *and* $\boldsymbol{FC}[V_1] \cup \boldsymbol{FC}[V_2] = \boldsymbol{FC}[V_1 + V_2]$*, the following holds:* $([[U_1]] = [[V_1]] \wedge [[U_2]] = [[V_2]]) \Rightarrow [[U_1 + U_2]] = [[V_1 + V_2]]$. *Such implication does not hold for semantics* $[[\ ]]_{par}$.

Proof Directly from (b) in Proposition 8.1. □

Proposition 8.5 *For arbitrary families* $\{U_j\}_{j \in I}$ *and* $\{V_j\}_{j \in I}$ *of c-e structures satisfying* $\bigcup_{j \in I} \boldsymbol{FC}[U_j] = \boldsymbol{FC}[\sum_{j \in I} U_j]$ *and* $\bigcup_{j \in I} \boldsymbol{FC}[V_j] = \boldsymbol{FC}[\sum_{j \in I} V_j]$*, where the least upper bounds involved exist, the following holds:* $\forall j \in I : [[U_j]] = [[V_j]] \Rightarrow [[\sum_{j \in I} U_j]] = [[\sum_{j \in I} V_j]]$. *Such implication does not hold for semantics* $[[\ ]]_{par}$.

Proof Directly follows from (b) and (d) in Proposition 8.3. □

8.3 Semantic Properties of Greatest Lower Bounds of Sets of c-e Structures

For the purpose of the forthcoming Theorem 8.1—the main fact on c-e structure semantics—we state some properties of the operation "$\circ$" introduced in Chap. 6 (Definition 6.1): $U \circ V = \sum(\boldsymbol{SUB}[U] \cap \boldsymbol{SUB}[V])$. Since $U \circ V = glb(U, V)$ and $U + V = \sum \boldsymbol{SUB}[U] \cup \boldsymbol{SUB}[V]) = lub(U, V)$, we obtain some properties dual to (c), (d) in Proposition 8.2.

Proposition 8.6 *For any c-e structures* $U, V \in \boldsymbol{CE}$:
(a) $[[U \circ V]] \subseteq [[U]] \cap [[V]]$
(b) $[[U \circ V]]^* \subseteq ([[U]] \cap [[V]])^* \subseteq [[U]]^* \cap [[V]]^*$
(c) None of the inclusions "$\subseteq$*" in (a) and (b) may be replaced with equality "*$=$*"*
(d) Properties (a)–(d) hold for semantics $[[\ ]]_{par}$.

Proof **Of (a)** directly follows from $U \circ V \leq U$, $U \circ V \leq V$ and (a) in Proposition 8.1.
Of (b) from (a) and general properties of relations.
Of (c) in (a): Let $U = \{a_x, b_{x \bullet y}, x^{a \bullet b}, y^b\}$, $V = \{a_y, b_{x \bullet y}, x^b, y^{a \bullet b}\}$, thus $U \circ V = \theta$, $[[U \circ V]] = \emptyset$, $[[U]] \cap [[V]] = [[U]] = [[V]] \neq \emptyset$. In (b)—evident by virtue of (a) and general properties of relations.
Of (d) –analogously. □

Remark Although $\boldsymbol{FC}[U] \cup \boldsymbol{FC}[V] = \boldsymbol{FC}[U + V] \Rightarrow [[U]] \cup [[V]] = [[U + V]]$ (Proposition 8.1 (b)), it is not true that $\boldsymbol{FC}[U] \cap \boldsymbol{FC}[V] = \boldsymbol{FC}[U \circ V] \Rightarrow [[U]] \cap [[V]] = [[U \circ V]]$, as shows the counterexample in the proof of (c) in Proposition 8.6. To find a structural condition for the semantic equation $[[U]] \cap [[V]] = [[U \circ V]]$, let us introduce a notion of ***closely connected*** firing component.

Definition 8.1 (*closely connected firing component—a completion*) A firing component $Q \in \boldsymbol{FC}$ is ***closely connected*** if
$\forall x \in {}^\bullet Q, \forall y \in Q^\bullet : Q = (x \to y) \bullet Q$, that is, there is an arrow from every cause $x \in {}^\bullet Q$ to every effect $y \in Q^\bullet$. For a given Q, by $\widehat{Q}$ is denoted the closely connected firing component with ${}^\bullet \widehat{Q} = {}^\bullet Q$ and $\widehat{Q}^\bullet = Q^\bullet$ called a ***completion*** of Q. The set $\widehat{G} = \{\widehat{Q} | \ Q \in G\}$, where $G \subseteq \boldsymbol{FC}$, is a *completion* of G. Obviously $[[Q]] = [[\widehat{Q}]]$ and $[[G]] = [[\widehat{G}]]$. □

Proposition 8.7 *For any c-e structures* $U, V \in \boldsymbol{CE}$:
(a) $\widehat{\boldsymbol{FC}[U]} \cup \widehat{\boldsymbol{FC}[V]} = \widehat{\boldsymbol{FC}[U + V]} \Rightarrow [[U]] \cup [[V]] = [[U + V]]$
(b) $\widehat{\boldsymbol{FC}[U]} \cap \widehat{\boldsymbol{FC}[V]} = \widehat{\boldsymbol{FC}[U \circ V]} \Rightarrow [[U]] \cap [[V]] = [[U \circ V]]$

Proof **Of (a)** By (c) in Proposition 8.2 it suffices to show $[[U + V]] \subseteq [[U]] \cup [[V]]$. If $(s, t) \in [[U + V]]$ then $(s, t) \in [[Q]]$ for a certain $Q \in \boldsymbol{FC}[U + V]$. Since $\widehat{Q} \in \widehat{\boldsymbol{FC}[U + V]} = \widehat{\boldsymbol{FC}[U]} \cup \widehat{\boldsymbol{FC}[V]}$ thus $\widehat{Q} \in \widehat{\boldsymbol{FC}[U]} \vee \widehat{Q} \in \widehat{\boldsymbol{FC}[V]}$ and $[[Q]] = [[\widehat{Q}]]$, thus $(s, t) \in [[\widehat{Q}]]$, hence $(s, t) \in \bigcup_{\widehat{Q} \in \widehat{\boldsymbol{FC}[U]}} [[\widehat{Q}]] \vee (s, t) \in \bigcup_{\widehat{Q} \in \widehat{\boldsymbol{FC}[V]}} [[\widehat{Q}]]$, thus $(s, t) \in \bigcup_{Q \in \boldsymbol{FC}[U]} [[Q]] \vee (s, t) \in \bigcup_{Q \in \boldsymbol{FC}[V]} [[Q]]$ (because $\bigcup_{\widehat{Q} \in \widehat{\boldsymbol{FC}[U]}} [[\widehat{Q}]] = \bigcup_{Q \in \boldsymbol{FC}[U]} [[Q]]$ and $\bigcup_{\widehat{Q} \in \widehat{\boldsymbol{FC}[V]}} [[\widehat{Q}]] = \bigcup_{Q \in \boldsymbol{FC}[V]} [[Q]]$). Therefore $(s, t) \in [[U]] \cup [[V]]$.
Of (b) By (a) in Proposition 8.6 it suffices to show, $[[U]] \cap [[V]] \subseteq [[U \circ V]]$. Let $(s, t) \in [[U]] \cap [[V]]$. Then $(s, t) \in [[U]] \wedge (s, t) \in [[V]]$ thus $(s, t) \in [[Q]] \wedge (s, t) \in [[P]]$, for some $Q \in \boldsymbol{FC}[U]$, $P \in \boldsymbol{FC}[V]$. Since $[[Q]] = [[\widehat{Q}]]$, $[[P]] = [[\widehat{P}]]$, we have $(s, t) \in [[\widehat{Q}]] \wedge (s, t) \in [[\widehat{P}]]$ thus $\widehat{Q} = \widehat{P}$. Hence $\widehat{Q} \in \widehat{\boldsymbol{FC}[U]} \wedge \widehat{Q} \in \widehat{\boldsymbol{FC}[V]}$, implying $\widehat{Q} \in \widehat{\boldsymbol{FC}[U]} \cap \widehat{\boldsymbol{FC}[V]}$ and by assumption, $\widehat{Q} \in \widehat{\boldsymbol{FC}[U \circ V]}$. Therefore $(s, t) \in [[U \circ V]]$. □

Proposition 8.8 (dual to Proposition 8.4) *For any c-e structures* $U_1, U_2, V_1, V_2 \in \boldsymbol{CE}$ *satisfying* $\widehat{\boldsymbol{FC}[U_1]} \cap \widehat{\boldsymbol{FC}[U_2]} = \widehat{\boldsymbol{FC}[U_1 \circ U_2]}$ *and* $\widehat{\boldsymbol{FC}[V_1]} \cap \widehat{\boldsymbol{FC}[V_2]} = \widehat{\boldsymbol{FC}[V_1 \circ V_2]}$, *the following holds: if* $[[U_1]] = [[V_1]]$ *and* $[[U_2]] = [[V_2]]$ *then* $[[U_1 \circ U_2]] = [[V_1 \circ V_2]]$. *Such implication does not hold for semantics* $[[\ ]]_{par}$.

Proof Directly from (b) in Proposition 8.7. □

8.4 Behavioural Equivalence and Conditional Congruence

Defining behavioural equivalence between c-e structures by $U \approx V$ if and only if $[[U]] = [[V]]$, we may rewrite Propositions 8.4 and 8.8 as:

if $U_1 \approx V_1$ and $U_2 \approx V_2$ then:

- $U_1 + U_2 \approx V_1 + V_2$ provided that $\boldsymbol{FC}[U_1] \cup \boldsymbol{FC}[U_2] = \boldsymbol{FC}[U_1 + U_2]$ and $\boldsymbol{FC}[V_1] \cup \boldsymbol{FC}[V_2] = \boldsymbol{FC}[V_1 + V_2]$
- $U_1 \circ U_2 \approx V_1 \circ V_2$ provided that $\widehat{\boldsymbol{FC}[U_1]} \cap \widehat{\boldsymbol{FC}[U_2]} = \widehat{\boldsymbol{FC}[U_1 \circ U_2]}$ and $\widehat{\boldsymbol{FC}[V_1]} \cap \widehat{\boldsymbol{FC}[V_2]} = \widehat{\boldsymbol{FC}[V_1 \circ V_2]}$

One may say that the equivalence $\approx \subseteq \boldsymbol{CE} \times \boldsymbol{CE}$ is a conditional congruence in the algebra $\langle \boldsymbol{CE}, +, \circ \rangle$. In the light of (b) in Proposition 8.1, the equivalence $\approx$ becomes a congruence for c-e structures with compositional semantics.

8.5 Conditional Homomorphism

As in Chaps. 2 and 3, by $\mathbb{S}$ is denoted the set of all states of c-e structures. Obviously, $(\mathbb{S} \times \mathbb{S}, \subseteq)$ or, equivalently, $(\mathbb{S} \times \mathbb{S}, \cup, \cap)$ is a lattice. It is a semantic domain of c-e structures. Let us introduce the following terminology. Let two algebraic systems $\mathscr{A} = \langle A, \{O_t\}_{t \in T} \rangle$, $\mathscr{A}' = \langle A', \{O_t^{'}\}_{t \in T} \rangle$, with the same arity of respective operations O_t and $O_t^{'}$, be given. Besides, let a condition Φ be given. A mapping $h\colon A \overset{into}{\longrightarrow} A'$ is a *Φ-conditional homomorphism* from $\mathscr{A}$ into $\mathscr{A}'$ when

$\Phi \Rightarrow h(O_t(a_1, a_2, ..., a_n)) = O_t^{'}(h(a_1), h(a_2), ..., h(a_n))$, for all $t \in T$.

Due to the results obtained in the above Propositions and terminology introduced, the main Theorem 8.1 may be stated.

Theorem 8.1 *Semantics* $[[\ ]]$ *of c-e structures is a Φ-conditional homomorphism from the lattice* $\langle \boldsymbol{CE}, +, \circ \rangle$ *into the lattice* $\langle \mathbb{S} \times \mathbb{S}, \cup, \cap \rangle$*, that is,*
$\Phi \Rightarrow [[U + V]] = [[U]] \cup [[V]] \wedge [[U \circ V]] = [[U]] \cap [[V]]$, *where* Φ *is:*
$\boldsymbol{FC}[U] \cup \boldsymbol{FC}[V] = \boldsymbol{FC}[U + V] \wedge \widehat{\boldsymbol{FC}[U]} \cap \widehat{\boldsymbol{FC}[V]} = \widehat{\boldsymbol{FC}[U \circ V]}$. □

The classic problems investigated in the Petri net theory, like reachability, liveness, boundedness and others, may be transfered onto the c-e structures, due to behavioural equivalence between the two descriptive systems. This equivalence is the subject of Chap. 9. Nonetheless, instead of such indirect obtainment of suitable results, let us examine the counterpart of liveness concept in Petri nets, more directly.

8.6 Soundness (Liveness) and Behavioural Deadlock

The structural deadlock has been introduced in Chap. 7 as a c-e structure U with $\boldsymbol{FC}[U] = \emptyset$. Such U can never exhibit any activity, regardless of the state. Now, we express some behavioural notions related to reaching states at which no action can be taken, as well as the opposite case: when such situation never can happen. Then, some of their properties will be proved. The considerations are restricted to c-e structures without self-loops, the counterpart of the pure Petri nets. A convenient way is to associate two predicates with a c-e structure U and state s:

$rec[U, s]$: $\mathbb{X} \to \{\textbf{true}, \textbf{false}\}$
$rid[U, s]$: $\mathbb{X} \to \{\textbf{true}, \textbf{false}\}$

defined as:
$rec[U, s](x) \stackrel{def}{=} \exists t : (s, t) \in [[U]] \wedge t(x) > s(x)$
(**true** iff the node x may receive some tokens at the state s)
$rid[U, s](x) \stackrel{def}{=} \exists t : (s, t) \in [[U]] \wedge s(x) > t(x)$
(**true** iff the node x may get rid of some tokens at the state s)
Let $alt[U, s](x) \stackrel{def}{=} rec[U, s](x) \vee rid[U, s](x)$.
The name "*alt*" stands for "alterable": $alt[U, s](x)$ says that node x may alter contents, that is change it at the state s. We restrict the considerations to semantics $[[\ ]]$. The reader is encouraged to examine properties stated below also for semantics $[[\ ]]_{par}$.

Proposition 8.9 *For any c-e structures U, V, any state s and any node x:*
(a) $alt[U, s](x) \vee alt[V, s](x) \Rightarrow alt[U + V, s](x)$ but reverse implication does not hold
(b) If $\boldsymbol{FC}[U] \cup \boldsymbol{FC}[V] = \boldsymbol{FC}[U + V]$ then $alt[U, s](x) \vee alt[V, s](x) \Leftrightarrow alt[U + V, s](x)$

Proof **Of (a).** If $alt[U, s](x) \vee alt[V, s](x)$ then either $rec[U, s](x) \vee rec[V, s](x)$ or $rid[U, s](x) \vee rid[V, s](x)$ holds. Let, for instance $rec[U, s](x) \vee rec[V, s](x)$. Then $[(s, t) \in [[U]] \wedge t(x) > s(x)] \vee [(s, t) \in [[V]] \wedge t(x) > s(x)]$, thus $(s, t) \in [[U]] \cup [[V]] \wedge t(x) > s(x)$ for a certain t. Hence, $(s, t) \in [[U + V]] \wedge t(x) > s(x)$ (by $[[U]] \cup [[V]] \subseteq [[U + V]]$—(c) in Proposition 8.2), implying $rec[U + V, s](x)$, thus $alt[U + V, s](x)$. Similarly for rid. The counter example to the reverse implication is: $U = \begin{pmatrix} a_{x \bullet y} & b_y \\ x^a & y^{a \bullet b} \end{pmatrix}$, $V = \begin{pmatrix} a_x & b_{x \bullet y} \\ x^{a \bullet b} & y^b \end{pmatrix}$, $s(a) = 1$ and remaining nodes of $\mathbb{X}$ do not hold tokens. Indeed, $U + V = \begin{pmatrix} a_{x \bullet y + x} & b_{x \bullet y + y} \\ x^{a \bullet b + a} & y^{a \bullet b + b} \end{pmatrix}$, $alt[U + V, s](a) = \textbf{true}$, but $alt[U, s](a) = \textbf{false}$ and $alt[V, s](a) = \textbf{false}$.
Of (b). Due to (a) we have only to verify the implication $alt[U + V, s](x) \Rightarrow alt[U, s](x) \vee alt[V, s](x)$ under assumption $\boldsymbol{FC}[U] \cup \boldsymbol{FC}[V] = \boldsymbol{FC}[U + V]$. By (b) in Proposition 8.1: $[[U + V]] = [[U]] \cup [[V]]$, thus, $rec[U + V, s](x) \Leftrightarrow \exists t : (s, t) \in [[U + V]] \wedge t(x) > s(x) \Leftrightarrow$

$\exists t : [(s,t) \in [[U]] \cup [[V]] \wedge t(x) > s(x)] \Rightarrow$
$\exists t : [(s,t) \in [[U]] \wedge t(x) > s(x) \vee \exists t : [(s,t) \in [[V]] \wedge t(x) > s(x)] \Leftrightarrow$
$rec[U,s](x) \vee rec[V,s](x)$. Similarly for rid. □

The predicate $alt[U,s]$ allows for defining the following types of state.

8.7 The Dead State for a Node

$dead[U,s](x) \stackrel{def}{=} \forall t : (s,t) \in [[U]]^* \Rightarrow \neg alt[U,t](x)$
In words: The state s is *dead* in U for a node x iff in no state, reachable from s, the node x can change its content.

8.8 The Ill State

$ill[U](s) \stackrel{def}{=} \exists x \in car(U) : dead[U,s](x)$
In words: The state s is *ill* in U iff s is dead for a certain $x \in car(U)$.

8.9 The Sound (Live) State

$sound[U](s) \stackrel{def}{=} \forall t : (s,t) \in [[U]]^* \Rightarrow \neg ill[U](t)$
In words: The state s is *sound* in U iff no *ill* state can be reached from s.

8.10 The Dead State

$dead[U](s) \stackrel{def}{=} \forall x \in car(U) : dead[U,s](x)$
In words: The state s is *dead* in U iff s is dead for each $x \in car(U)$.

Note that if a c-e structure U is a structural deadlock then each state is dead in U. Note also that the concept of *sound* state is a counterpart to the notion of *live* marking in Petri nets.

Let us investigate how the *deadness*, *illness*, *soundness* for a c-e structure, affects its substructures and conversely. Recall that $\sum \boldsymbol{SUB}[W] = W$ (see (b) in Proposition 6.1). We limit ourselves to two constituents U, V of W: $U + V = W$ but the results may be easily extended to any, also to infinite number of constituents. In the latter case, the upper bounds must be involved.

Proposition 8.10 *For any c-e structures U, V, any state s and node x:*
(a) $dead[U + V, s](x) \Rightarrow dead[U, s](x) \wedge dead[V, s](x)$ *but the reverse implication not always holds.*
(b) If $\boldsymbol{FC}[U] \cup \boldsymbol{FC}[V] = \boldsymbol{FC}[U + V]$ *then* $dead[U + V, s](x) \Leftrightarrow dead[U, s](x) \wedge dead[V, s](x)$
(c) $ill[U + V](s) \Rightarrow ill[U](s) \wedge ill[V](s)$ *but the reverse implication not always holds.*
(d) If $\boldsymbol{FC}[U] \cup \boldsymbol{FC}[V] = \boldsymbol{FC}[U + V]$ *then* $ill[U + V](s) \Leftrightarrow ill[U](s) \wedge ill[V](s)$

Proof **Of (a)**. Suppose $\neg(dead[U, s](x) \wedge dead[V, s](x))$ that is, $\neg dead[U, s](x) \vee \neg dead[V, s](x)$. Let, for instance, $\neg dead[U, s](x)$. Thus, $\exists t : ((s, t) \in [[U]]^* \wedge alt[U, t](x))$, which implies $\exists t : ((s, t) \in [[U + V]]^* \wedge alt[U + V, t](x))$, because $[[U]]^* \subseteq [[U + V]]^*$ and, by (a) in Proposition 8.9, $alt[U, t](x) \Rightarrow alt[U + V, t](x)$. Hence, $\neg dead[U + V, s](x)$ holds. The same is obtained if $\neg dead[V, s](x)$ holds. The counter example for the reverse implication is the same as in Proposition 8.9(a).
Of (b) $dead[U + V, s](x) \Leftrightarrow \forall t : ((s, t) \in [[U + V]]^* \Rightarrow \neg alt[U + V, t](x)) \Leftrightarrow$ (by (b) in Proposition 8.1) $\forall t : ((s, t) \in ([[U]] \cup [[V]])^* \Rightarrow \neg alt[U + V, t](x)) \Rightarrow$ (by (h) in Proposition 2.3) $\forall t : ((s, t) \in [[U]]^* \cup [[V]]^* \Rightarrow \neg alt[U + V, t](x)) \Rightarrow$ (by (b) in Proposition 8.9) $\forall t : ((s, t) \in [[U]]^* \cup [[V]]^* \Rightarrow \neg alt[U, t](x) \wedge \neg alt[V, t](x)) \Leftrightarrow \forall t : (((s, t) \in [[U]]^* \vee (s, t) \in [[V]]^*) \Rightarrow (\neg alt[U, t](x) \wedge \neg alt[V, t](x)))$. This implies: $\forall t : ((s, t) \in [[U]]^* \Rightarrow \neg alt[U, t](x))$ and $\forall t : ((s, t) \in [[V]]^* \Rightarrow \neg alt[V, t](x))$, hence $dead[U, s](x) \wedge dead[V, s](x)$ holds.
Of (c) and **(d)**—follow from (a) and (b) of this proposition.

Propositions 8.9 and 8.10 account for some relationships between *alterable/dead/ill* states in the composite c-e structure $W = U + V$ and in its constituents U, V. Are there similar relationships between *sound/dead* states in W and in U, V? The answer is no, as states the Theorem 8.2.

Theorem 8.2 *Let* $\diamond$ *denote either* $\vee$ *or* $\wedge$. *There exist c-e structures U, V and a state s, such that:*
(a) neither $sound[U](s) \diamond sound[V](s) \Rightarrow sound[U + V](s)$
nor $sound[U + V](s) \Rightarrow sound[U](s) \diamond sound[V](s)$
(b) neither $dead[U](s) \diamond dead[V](s) \Rightarrow dead[U + V](s)$
nor $dead[U + V](s) \Rightarrow dead[U](s) \diamond dead[V](s)$

Proof By the following counter examples:
Of (a). Let $U = \begin{pmatrix} a_{x+y}^{x \bullet y} & b_{x \bullet y}^{x \bullet y} \\ x_{a+b}^{a \bullet b} & y_{a \bullet b}^{a \bullet b} \end{pmatrix}$ $V = \begin{pmatrix} a_{x \bullet y}^{x+y} & b_{x \bullet y}^{x \bullet y} \\ x_{a \bullet b}^{a+b} & y_{a \bullet b}^{a \bullet b} \end{pmatrix}$ with $s(a) = s(b) = 1$.
Then, $U + V = \begin{pmatrix} a_{x \bullet y+x+y}^{x \bullet y+x+y} & b_{x \bullet y}^{x \bullet y} \\ x_{a \bullet b+a+b}^{a \bullet b+a+b} & y_{a \bullet b}^{a \bullet b} \end{pmatrix}$. Thus $sound[U](s) =$ **false** and $sound[V](s) =$ **false** but $sound[U + V](s) =$ **true**. Hence implication $sound[U + V](s) \Rightarrow sound[U]$
$(s) \diamond sound[V](s)$ does not hold. On the other hand, consider U, V and $U + V$ with $s(a) = s(b) = 1$ depicted in Fig. 8.2. Thus, Fig. 8.2 shows that $sound[U](s) =$ **true**

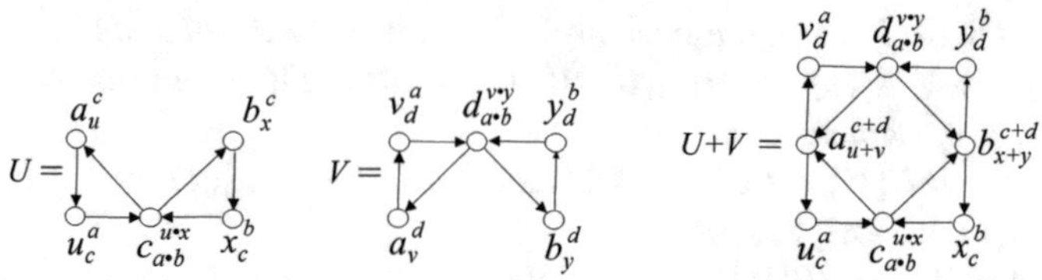

Fig. 8.2 The state s, with $s(a) = s(b) = 1$, is sound (live) in U and V but not in $U + V$

and $sound[V](s) = \mathbf{true}$ but $sound[U + V](s) = \mathbf{false}$, because state t with $t(v) = t(x) = 1$ is reachable from s in $U + V$ and $ill[U + V](t) = \mathbf{true}$ (no move at the state t is possible—a total deadlock state!). The same concerns state t' with $t'(u) = t'(y) = 1$, reachable from s.

Of (b). Use the same counterexamples as for $sound$. □

Chapter 9
Relation of Cause-Effect Structures to Petri Nets

At the first glance it seems, that transformation of c-e structures into Petri nets [1] and vice versa should be straightforward: the c-e structure's nodes play role of net places whereas firing components—transitions. Such simple 1-1 correspondence, however, not always is possible to be established from nets to c-e structures. This fact was first noted by M. Raczunas in [2]. Nonetheless, the 1-1 correspondence preserving behavioural equivalence between c-e structures and Petri nets is possible, following a slight modification of the c-e structures—if needed. The two-way transformation (c-e structures $\longleftrightarrow$ Petri nets) dealt with, concerns the structure, thus, for simplicity, the considerations are restricted to elementary c-e structures and Petri nets. Thus, regard of weights of edges and capacity of places, would not bring anything new to the considerations. Before going into details let us remind the notation for the structure (schema) of such nets.

9.1 Elementary Petri Nets

A net (schema) is a triple $\mathcal{N} = \langle S, T, F\rangle$ where S is a set of places, T—a set of transitions with $S \cap T = \emptyset$, and $F \subseteq S \times T \cup T \times S$ set of directed edges, called a flow relation. A *pre-set* and *post-set* of a transition $t \in T$ is ${}^\bullet t = \{x \in S|\ (x, t) \in F\}$ and $t^\bullet = \{x \in S|\ (t, x) \in F\}$ respectively and dually for places $x \in S$: $t \in {}^\bullet x \Leftrightarrow x \in t^\bullet$ and $t \in x^\bullet \Leftrightarrow x \in {}^\bullet t$. To conform to the construction of c-e structures, let us admit (a not essential) restrictions: ${}^\bullet t \neq \emptyset \neq t^\bullet$ and $t_1 \neq t_2 \Rightarrow ({}^\bullet t_1 \neq {}^\bullet t_2 \vee t_1^\bullet \neq t_2^\bullet)$ for any $t, t_1, t_2 \in T$. The latter means that we consider the so-called simple nets only. A marking M of a net is a counterpart of the c-e structures' state and firing rule—a counterpart of semantics: $M \subseteq S$ and $(M, M') \in [[t]]$ iff ${}^\bullet t \subseteq M \wedge t^\bullet \cap M = \emptyset \wedge M' = (M \setminus {}^\bullet t) \cup t^\bullet$ for $t \in T$ and $[[\mathcal{N}]] = \bigcup\limits_{t \in T} [[t]]$. Now, let us define a strong equivalence between the c-e structures and nets and the direct transformations between them.

L. Czaja, *Cause-Effect Structures*, Lecture Notes in Networks and Systems 45, https://doi.org/10.1007/978-3-030-20461-7_9

Definition 9.1 (*strong equivalence*) A net $\mathcal{N} = \langle S, T, F \rangle$ and a c-e structure $U = \langle C, E \rangle$ are *strongly equivalent* if and only if there exist two bijections $g : S \to car(U)$ and $f : T \to \boldsymbol{FC}[U]$ such that $g(^\bullet t) = {}^\bullet f(t)$ and $g(t^\bullet) = f(t)^\bullet$ for each $t \in T$, where $g(^\bullet t) = \{g(x)|\ x \in {}^\bullet t\}$, $g(t^\bullet) = \{g(x)|\ x \in t^\bullet\}$. □

Obviously, ${}^\bullet f(t)$ and $f(t)^\bullet$ are pre and post sets of the firing component $f(t)$—see Definition 2.5 in Chap. 2, while ${}^\bullet t$ and $t^\bullet$ are pre and post sets of the transition t. This structural equivalence yields the identical behaviour: if a net $\mathcal{N}$ and c-e structure U are strongly equivalent then $[[\mathcal{N}]]$ and $[[U]]$ are identical up to naming of places in $\mathcal{N}$ and nodes in U.

Definition 9.2 (*direct transformation*) Suppose that places of nets and nodes in c-e structures are members of the same universe $\mathbb{X}$. For a given c-e structure $U = \langle C, E \rangle$ define a net $\mathcal{N} = \langle S, T, F \rangle$ as follows: $S = car(U)$, $T = \boldsymbol{FC}[U]$, $F = \{(x, Q)|\ x \in {}^\bullet Q \wedge Q \in T\} \cup \{(Q, x)|\ x \in Q^\bullet \wedge Q \in T\}$. For a given net $\mathcal{N} = \langle S, T, F \rangle$ define a c-e structure $U = \langle C, E \rangle$ as follows:

$$C(x) = \sum_{t \in T_x} \bigwedge_{y \in {}^\bullet t} y \qquad \text{where } T_x = \{t \in T|\ x \in t^\bullet\}$$

$$E(x) = \sum_{t \in T^x} \bigwedge_{y \in t^\bullet} y \qquad \text{where } T^x = \{t \in T|\ x \in {}^\bullet t\}$$

(As in Chap. 2, "$\bigwedge$" stands for the operator of repetitive multiplication "•" of c-e structures).

These transformations are called *direct*. □

For example, if $U = \begin{pmatrix} a_{x\bullet y+y} & b_y \\ x^a & y^{a\bullet b+a} \end{pmatrix}$ then the net obtained by the direct transformation of U is depicted in Fig. 9.1, where the shaded boxes—transitions are the U's firing components. For illustration, the latter are put in the boxes.

The direct transformation of any c-e structure yields a strongly equivalent net: the mappings $g(x) = x$ and $f(Q) = Q$, for $x \in car(U)$, $Q \in \boldsymbol{FC}[U]$ satisfy requirements in Definition 9.1 when the net $\mathcal{N}$ is constructed from U according to Definition 9.2. Let us consider the converse situation. In Fig. 9.2 a net and its strongly equivalent direct transformation is depicted.

As in Definition 9.2, we assume that net-places and nodes of c-e structures are taken from the set $\mathbb{X}$. However not always a strongly equivalent c-e structure exists for a given net, as the example in Fig. 9.3 demonstrates.

Indeed, the left transition in Fig. 9.3 transforms into firing component $\{a_{x\bullet y}, x^a, y^a\}$ and the right—into $\{a_y, b_y, y^{a\bullet b}\}$. Their sum, that is c-e structure obtained by the direct transformation, contains two more firing components: $\{a_{x\bullet y}, b_y, x^a, y^{a\bullet b}\}$ and $\{a_y, y^a\}$. The Petri net and the c-e structure behave differently. Definition 9.2 prompts a simple algorithm for the direct transformation of nets into c-e structures, thus not always preserving behavioural equivalence.

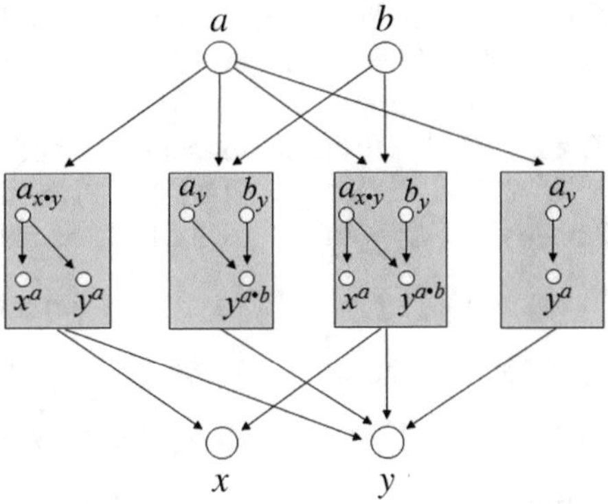

Fig. 9.1 The Petri net strongly equivalent to c-e structure $U = \{a_{x\bullet y+y}, b_y, x^a, y^{a\bullet b+a}\}$

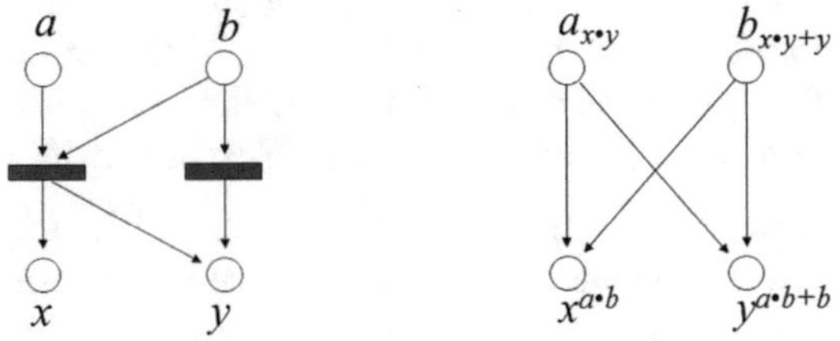

Fig. 9.2 The Petri net and its strongly equivalent direct transformation

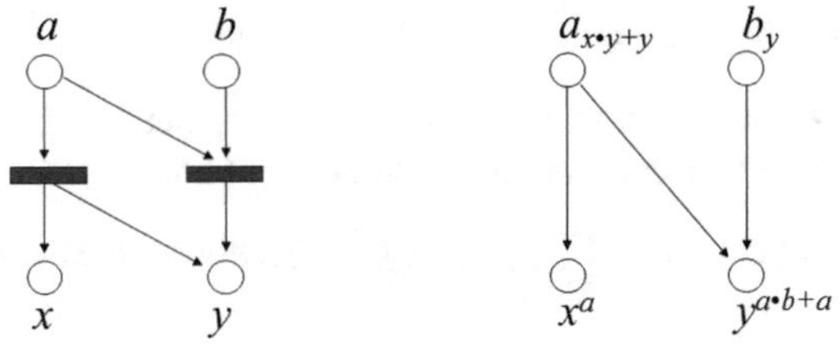

Fig. 9.3 The Petri net and its direct transformation. This net has no strongly equivalent c-e structure

9.2 Algorithm for the Direct Transformation of Petri Nets onto c-e Structures

Let a Petri net $\mathcal{N} = \langle S, T, F \rangle$ be given. For any transition $t \in T$ construct a firing component Q_t as follows: ${}^\bullet Q_t = {}^\bullet t$, $Q_t^\bullet = t^\bullet$ and to each node $x \in {}^\bullet Q_t$ assign $\bigwedge_{t \in T} Q_t^\bullet$ as the subscript and to each node $x \in Q_t^\bullet$ assign $\bigwedge_{t \in T} {}^\bullet Q_t$ as the superscript. Sum up all the firing components thus constructed. Resulting c-e structure is identical with U obtained by the direct transformation in Definition 9.2.

The following Theorem 9.1 is a criterion of existence of a strongly equivalent c-e structure for a given Petri net.

Theorem 9.1 *There exists a c-e structure strongly equivalent to a given net if and only if the c-e structure constructed by above algorithm of direct transformation, is strongly equivalent.*

Proof Suppose there exists a c-e structure V strongly equivalent to $\mathcal{N} = \langle S, T, F \rangle$. Thus $|\boldsymbol{FC}[V]| = |T|$. Assuming $car(V) = S$ (otherwise, rename suitably nodes in V to make it the case), we get ${}^{\bullet}t = {}^{\bullet}P_t$ and $t^{\bullet} = P_t^{\bullet}$ for each $t \in T$, where $P_t \in \boldsymbol{FC}[V]$ is a firing component corresponding to transition t by strong equivalence of $\mathcal{N}$ and V. Since $\sum \boldsymbol{FC}[V] \leq V$ ((b) in Proposition 6.3), by (c) in Proposition 2.3 we get $\boldsymbol{FC}[\sum \boldsymbol{FC}[V]] \subseteq \boldsymbol{FC}[V]$, hence $\boldsymbol{FC}[\sum \bigcup_{t \in T}\{P_t\}] = \boldsymbol{FC}[\sum_{t \in T} P_t] \subseteq \boldsymbol{FC}[V]$ (because $\boldsymbol{FC}[V] = \bigcup_{t \in T}\{P_t\}$). On the other hand, by Proposition 6.3(c): $\bigcup_{t \in T} \boldsymbol{FC}[P_t] \subseteq \boldsymbol{FC}[\sum_{t \in T} P_t]$, hence $\bigcup_{t \in T}\{P_t\} \subseteq \boldsymbol{FC}[\sum_{t \in T} P_t]$ (because $\boldsymbol{FC}[P_t] = \{P_t\}$), that is $\boldsymbol{FC}[V] \subseteq \boldsymbol{FC}[\sum_{t \in T} P_t]$. Therefore $\boldsymbol{FC}[\sum_{t \in T} P_t] = \boldsymbol{FC}[V]$, which means that summation of firing components P_t does not create new firing components. Let Q_t be a firing component constructed from transition t by the above algorithm of direct transformation. Obviously ${}^{\bullet}Q_t = {}^{\bullet}t = {}^{\bullet}P_t$ and $Q_t^{\bullet} = t^{\bullet} = P_t^{\bullet}$ and—by construction—Q_t is closely connected firing component (Definition 8.1), meaning: there is an arrow from each $x \in {}^{\bullet}Q_t$ to each $y \in Q_t^{\bullet}$. Now, it remains to note that $\boldsymbol{FC}[\sum_{t \in T} P_t] = \boldsymbol{FC}[V]$ implies strong equivalence between U and $\mathcal{N}$. Replace any firing component P_t in V with its completion $\widehat{P_t}$, that is closely connected firing component with the same as in P_t pre and post set and denote the resulting c-e structure by $\widehat{V}$. For example, if $V = \{a_{x \bullet y+y},\ b_y,\ x^a,\ y^{a \bullet b+a}\}$ then $\widehat{V} = \{a_{x \bullet y+y},\ b_{x \bullet y+y},\ x^{a \bullet b+a},\ y^{a \bullet b+a}\}$. Obviously $\widehat{V}$ has only $\widehat{P_t}$ $(t \in T)$ as its firing components, thus $\boldsymbol{FC}[\sum_{t \in T} \widehat{P_t}] = \boldsymbol{FC}[\widehat{V}]$, hence $\widehat{V}$ is strongly equivalent with the net $\mathcal{N}$. Since $\widehat{P_t} = Q_t$ and (by construction of U by the above algorithm of direct transformation) $U = \sum_{t \in T} Q_t$, we get $\boldsymbol{FC}[U] = \boldsymbol{FC}[\sum_{t \in T} Q_t] = \boldsymbol{FC}[\widehat{V}]$. Therefore U is strongly equivalent to $\mathcal{N}$. □

The reason for non-existence of a strongly equivalent c-e structure for each Petri net is, obviously, the creation of superfluous (redundant) firing components in summation—recall Proposition 2.2, points (m) and (n). Although not always is it possible to construct a strongly equivalent c-e structure for a given net, this news is not as bad as it might seem. An equivalence, no matter whether defined structurally or semantically, should reflect a similarity of behaviour. In the case of strong equivalence, the similarity is complete: not only the Petri net and its c-e structure counterpart generate the same (up to 1-1 renaming) strings of transitions and firing components, but also the same markings and states. The requirement of such similarity is usually too demanding in the system design practice. Therefore, we shall relax this restrictive equivalence, retaining the capability of generating the same strings of transitions and firing components only. To this end, let us define an auxiliary equivalence relation between net's places as well as between c-e structure's nodes. This will allow for grouping places (nodes) into equivalence classes. All the members of a class are intended to simultaneously hold or not to hold tokens during all the c-e structure's (to be constructed from a given net) activity.

Definition 9.3 (*equivalence "≡" of Petri net places or c-e structure's nodes*) Let a net $\mathcal{N} = \langle S, T, F\rangle$ and a c-e structure U be given. For $x, y \in S$ or $x, y \in car(U)$ define $x \equiv y$ if and only if ${}^\bullet x = {}^\bullet y$ and $x^\bullet = y^\bullet$. Obviously $\equiv$ is an equivalence relation. Let us shorten the denotation of $\equiv$-equivalence classes $[x]_\equiv$ to $[x]$ and for a set $A \subseteq S$ or $A \subseteq car(U)$ let us denote the set $\{[x] : x \in A\}$ by $[A]$. □

Remark (1) At each marking reachable from an initial one, all the places (nodes) in the class $[x]$, either simultaneously hold tokens or they do not hold tokens—provided that this is the case at the initial marking.
(2) $[A]$ is the quotient set (with respect to $\equiv$) of A, that is the set of all equivalence classes of the A's members: $[A] = A/_\equiv$.

We can define a weaker (than strong) equivalence between Petri nets and c-e structures.

Definition 9.4 (*weak equivalence*) A net $\mathcal{N} = \langle S, T, F\rangle$ and a c-e structure $U = \langle C, E\rangle$ are *weakly equivalent* if and only if there are two bijections $g : [S] \to [car(U)]$ and $f : T \to \boldsymbol{FC}[U]$ such that $g([{}^\bullet t]) = [{}^\bullet f(t)]$ and $g([t^\bullet]) = [f(t)^\bullet]$ for each $t \in T$. □

Thus, the difference between strong and weak equivalence concerns the bijection g only: in the former case its domain and codomain are sets S and $car(U)$, whereas in the latter—their quotients. It is seen that this definition ensures the property expressed in point (1) of the above Remark. That is, a transition $t \in T$ is firable iff all the places $x \in \bigcup[{}^\bullet t]$ hold tokens and all the places $y \in \bigcup[t^\bullet]$—do not.

Now, to construct a transformation of nets into weakly equivalent c-e structures we take advantage of a characteristic feature of firing components: there are, usually, many of them with common behaviour, that is with the same pre-set and the same post-set. For example, the following five firing components behave identically: $\{a_{x\bullet y}, b_{x\bullet y}, x^{a\bullet b}, y^{a\bullet b}\}$, $\{a_{x\bullet y}, b_y, x^a, y^{a\bullet b}\}$, $\{a_x, b_{x\bullet y}, x^{a\bullet b}, y^b\}$, $\{a_{x\bullet y}, b_x, x^{a\bullet b}, y^a\}$, $\{a_y, b_{x\bullet y}, x^b, y^{a\bullet b}\}$.

The key idea of the transformation is to replicate—if needed—some places to obtain groups of $\equiv$-equivalent places. The replication should go as long as the replacement of closely connected firing components (obtained by the direct transformation of transitions) by some of their not closely connected but behaviourally equivalent counterparts, makes all the superfluous firing components (created in summation) disappear.

Before presenting the algorithm, let us illustrate this procedure on example of the net in Fig. 9.4. Firing components $Q_t = \{a_{x\bullet y}, x^a, y^a\}$ and $Q_u = \{a_y, b_y, y^{a\bullet b}\}$ are direct transformations of transitions t and u. C-e structure $Q_t + Q_u$ contains, apart from Q_t and Q_u, two superfluous firing components $\{a_{x\bullet y}, b_y, x^a, y^{a\bullet b}\}$ and $\{a_y, y^a\}$. We eliminate them as follows. First, replicate node a into the group $\{a[t], a[u]\}$ and y into $\{y[t], y[u]\}$ and convert Q_t and Q_u respectively into:

$$Q'_t = \{a[t]_{x\bullet y[t]\bullet y[u]}, a[u]_{x\bullet y[t]\bullet y[u]}, x^{a[t]\bullet a[u]}, y[t]^{a[t]\bullet a[u]}, y[u]^{a[t]\bullet a[u]}\}$$

$$Q'_u = \{a[t]_{y[t]\bullet y[u]}, a[u]_{y[t]\bullet y[u]}, b_{y[t]\bullet y[u]}, y[t]^{a[t]\bullet a[u]\bullet b}, y[u]^{a[t]\bullet a[u]\bullet b}\}$$

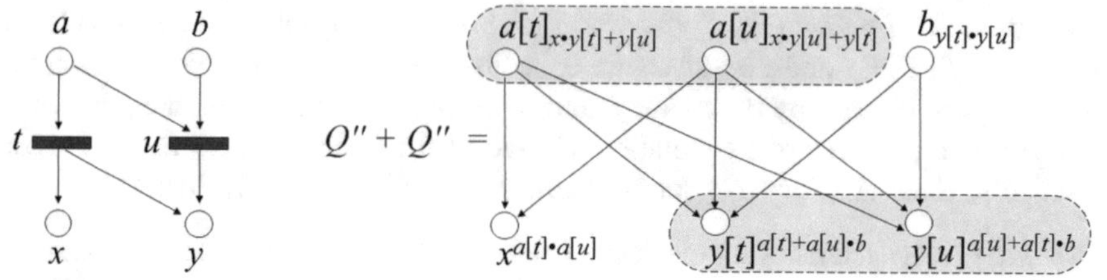

Fig. 9.4 Petri net and its weakly equivalent c-e structure. Groups of the replicas of a and y are encircled and shaded

Obviously, Q'_t, Q'_u behave like Q_t, Q_u if the replicas $a[t]$, $a[u]$ and $y[t]$, $y[u]$ are marked with tokens like their originals a and y. Second, replace Q'_t and Q'_u by their behaviourally equivalent (but not closely conected) counterparts Q''_t and Q'':

$$Q''_t = \{a[t]_{x\bullet y[t]},\ a[u]_{x\bullet y[u]},\ x^{a[t]\bullet a[u]},\ y[t]^{a[t]},\ y[u]^{a[u]}\}$$

$$Q''_u = \{a[t]_{y[u]},\ a[u]_{y[t]},\ b_{y[t]\bullet y[u]},\ y[t]^{a[u]\bullet b},\ y[u]^{a[t]\bullet b}\}$$

It is seen that c-e structure $Q''_t + Q''_u$ depicted in Fig. 9.4 contains Q''_t and Q''_u only as its firing components: the superfluous ones disappeared! Q''_t and Q''_u are obtained by suitable deletion of some arguments in the lower and upper monomials in Q'_t and Q'_u. Note also that $a[t] \equiv a[u]$ and $y[t] \equiv y[u]$ in the c-e structure $Q''_t + Q''_u$ (and also in $Q'_t + Q'_u$).

This example illustrates the algorithm of transforming Petri nets into weakly equivalent c-e structures. For its better understanding, note that the creation of superfluous firing components in summation stems from combining some polynomials present in distinct summands into new polynomials present in the sum. In the above example, for instance, monomials $x \bullet y$ and a in $Q_t + Q_u = \{a_{x\bullet y+y},\ b_y,\ x^a,\ y^{a\bullet b+a}\}$ come from Q_t, while monomials $a \bullet b$ and y—from Q_u. When combined into new polynomials $x \bullet y + y$ and $a \bullet b + a$, the track where the monomials come from is lost. This causes creation of superfluous firing components. For keeping the track one may parametrize each node in the summands by a name of the summand the node belongs to. In this way some nodes are replicated and new summands formed. Then, by replacing some of them with their suitably chosen (behavioural) equivalents, one can eliminate the superfluous firing components. This procedure may be presented as follows.

Algorithm of Transformation of Petri Nets into Cause-Effect Structures

Input: net $\mathcal{N} = \langle S, T, F\rangle$
Output: c-e structure $U = \langle C_U, E_U\rangle$ equivalent (strongly or weakly) to $\mathcal{N}$.

Step 1 (check of existence of a c-e structure strongly equivalent to $\mathcal{N}$)
Transform directly $\mathcal{N}$ into a c-e structure $V = \langle C_V, E_V\rangle$, using the above **Algorithm for the direct transformation of nets onto c-e structures**. If $|T| = |FC[V]|$ then define $U = V$ and stop (U is a strongly equivalent to $\mathcal{N}$), else go to Step 2.

Step 2 (replication of nodes)
In the c-e structure V obtained in Step 1 replace each node $x \in S = car(V)$ by the group $\{x[t] : t \in {}^{\bullet}x \cup x^{\bullet}\}$ of its replicas and each argument y in polynomials $C_V(x)$, $E_V(x)$—by the product of the y's replicas (that is members of the group $\{y[t] : t \in {}^{\bullet}y \cup y^{\bullet}\}$). *Comment*: all the replicas of any node are $\equiv$-equivalent. Denote by V' c-e structure obtained from V in such way and go to Step 3.

Step 3 (elimination of superfluous firing components)
Decompose the c-e structure V' obtained in Step 2 into firing components. Partition their set $\boldsymbol{FC}[V']$ into those corresponding to transitions $t \in T$, denote their set by $\boldsymbol{FC}_1[V']$, and the remaining, that is superfluous firing components collected in the set denoted by $\boldsymbol{FC}_2[V']$. Obviously, each firing component $Q'_t \in \boldsymbol{FC}_1[V']$ is closely connected: $Q'_t = \widehat{Q'_t}$. By deleting some arguments in the polynomials, convert each $Q'_t \in \boldsymbol{FC}_1[V']$ into a behaviourally equivalent Q''_t (that is such that ${}^{\bullet}Q''_t = {}^{\bullet}Q'_t$, $Q''^{\bullet}_t = Q'^{\bullet}_t$) so that the sum $\sum_{t \in T} Q''_t$ does not contain superfluous firing components: $\boldsymbol{FC}[\sum_{t \in T} Q''_t] = \{Q''_t \mid t \in T\}$. Existence of such firing components Q''_t—not closely connected—is the main result of the paper [2]. Define $U = \sum_{t \in T} Q''_t$ (U is weakly equivalent to $\mathcal{N}$) and stop.

End of the Algorithm

Remark (1) If the set $\{x[t] : t \in {}^{\bullet}x \cup x^{\bullet}\}$ of the x's replicas defined in Step 2 contains one member only, we may leave x not renamed—as has been done in the example preceding the algorithm.
(2) A somewhat different algorithm of transforming nets into c-e structures has been implemented by Rokoszewski [3].
(3) Let $U_j = \langle C_{U_j}, E_{U_j} \rangle$, $j = 1, 2, ..., n$ be c-e structures. A sufficient (but not necessary) condition for a sum $U = \sum_{j=1}^{n} U_j$ to contain firing components from its summands only, that is $\boldsymbol{FC}[U] = \bigcup_{j=1}^{n} \boldsymbol{FC}[U_j]$, is the following: for each $x \in \bigcup_{j=1}^{n} car(U_j)$ either the set of arguments occurring in $C_{U_i}(x)$ is disjoint with the set of arguments in $C_{U_j}(x)$ when $i \neq j$ or when the same holds for $E_{U_i}(x)$ and $E_{U_j}(x)$. This condition may serve as a guideline in deleting some arguments from the polynomials in the c-e structures (firing components) Q'_t—in Step 3.

References

1. Petri CA (1966) Communication with automata. Final report, vol 1, Supplement 1, RADC TR-65-337-vol1-suppl 1. Applied Research, Princeton, NJ. Contract AF 30(602)-3324
2. Raczunas M (1993) Remarks on the equivalence of c-e structures and Petri nets. Inf Proc Lett 45:165–169
3. Rokoszewski R (2000) A system of testing equivalence between Petri nets and cause-effect structures: a project and program. MSc thesis (in Polish), Warsaw University

Chapter 10
Processes in Elementary Cause-Effect Structures

When a firing component is executed (“fired”), an event occurs. Events are actions of “fire off” of a certain set of firing components, in particular during performance of a c-e structure. They yield change of state. Formally, an event is a firing component whose causes and effects are accompanied by their occurrence number. More accurately, if a firing component from a certain set is Q then its action results in event represented by Q in which each node $x \in {}^{\bullet}Q^{\bullet}$ is replaced with a pair $< x, k >$, where k is a number of x's occurrences on a certain acting stage of the firing components from this set. Event $< x, k >$ is called the k-th *occurrence* of x. A process will be defined as the sum ($\sum$) of some events, their least upper bound in general. Thus, processes are also c-e structures, but of a limited form (acyclic and with no “+” in polynomials assigned to nodes) and over the set of events from the set $\mathbb{Y} = \mathbb{X} \times \mathbb{N}$ ($\mathbb{N}$—the set of natural numbers). It will be shown that the numbering of node occurrences determines a precedence relation between events (their partial ordering) in a process, thus also their concurrency. Therefore, a process in a c-e structure represents its particular activity (run), recorded as a set of events partially ordered by the precedence relation. Examples are in Figs. 1.4, 10.1 and 10.4. We shall explicitly indicate the set $\mathbb{X}$ or $\mathbb{Y}$ over which some constructs, like sets of sequences of firing components or processes, are built up. As usually $\boldsymbol{FC}^{*}_{\mathbb{X}}$ and $\boldsymbol{FC}^{\omega}_{\mathbb{X}}$ denote respectively, the set of all finite and infinite strings of firing components over $\mathbb{X}$. In case of $\boldsymbol{FC}^{*}_{\mathbb{X}}$, the empty string ε is included. For $u = Q_1Q_2Q_3 \ldots \in \boldsymbol{FC}^{*}_{\mathbb{X}} \cup \boldsymbol{FC}^{\omega}_{\mathbb{X}}$, by $|u|$ is denoted the length of the sequence u, i.e. number of its members if it is finite or ∞—if infinite.

Definition 10.1 (*firing sequence, u-computation*) A string $u = Q_1Q_2Q_3 \ldots \in \boldsymbol{FC}^{*} \cup \boldsymbol{FC}^{\omega}_{\mathbb{X}}$, is a *firing sequence* if $u = \varepsilon$ or there are states $s_0, s_1, s_2, \ldots$ with $(s_{j-1}, s_j) \in [[Q_j]]$, $j \geq 1$. This sequence of states is a *u-computation*. The set of all firing sequences over $\mathbb{X}$ is denoted by $\boldsymbol{FS}_{\mathbb{X}}$ (thus $\boldsymbol{FS}_{\mathbb{X}} \subset \boldsymbol{FC}^{*}_{\mathbb{X}} \cup \boldsymbol{FC}^{\omega}_{\mathbb{X}}$), its members denoted by $u, v, w, \ldots$. □

L. Czaja, *Cause-Effect Structures*, Lecture Notes in Networks and Systems 45, https://doi.org/10.1007/978-3-030-20461-7_10

This definition, by the inductive argument implies:

Proposition 10.1 *A string* $u = Q_1Q_2Q_3\ldots \in \boldsymbol{FC}_{\mathbb{X}}^{*} \cup \boldsymbol{FC}_{\mathbb{X}}^{\omega}$, *is a firing sequence iff each finite prefix of* u *is a firing sequence.*

□

Recall that in Chap. 2, by $\boldsymbol{CE}[\mathbb{X}]$ (abbreviated to $\boldsymbol{CE}$) is denoted the set of all c-e structures over the space of nodes $\mathbb{X}$. Since processes will be c-e structures over the space of nodes $\mathbb{Y} = \mathbb{X} \times \mathbb{N}$, they belong to $\boldsymbol{CE}[\mathbb{Y}]$. Although the least upper bound $\sum \boldsymbol{CE}[\mathbb{Y}]$ does not exist (point (d) in Proposition 6.2 in Chap. 6), let us introduce an artificial element $\Omega_{\mathbb{Y}}$ complementing the set $\boldsymbol{CE}[\mathbb{Y}]$ and satisfying $V \leq \Omega_{\mathbb{Y}}$ for all $V \in \boldsymbol{CE}[\mathbb{Y}]$. Thus, $V + \Omega_{\mathbb{Y}} = \Omega_{\mathbb{Y}}$, i.e. $\Omega_{\mathbb{Y}}$ absorbs in summation any c-e structure over $\mathbb{Y}$. That is, $\Omega_{\mathbb{Y}}$ may be regarded as symbolising "*chaos*"—an unstructured c-e structure. Hence, for any c-e structure $V \in \boldsymbol{CE}[\mathbb{Y}]$: $\theta \leq V \leq \Omega_{\mathbb{Y}}$. For the purpose of process definition, let $u_j(x)$ denote number of occurrences of the node x within elements of the prefix $u_j = Q_1Q_2\ldots Q_j$ of the firing sequence $u = Q_1Q_2\ldots Q_jQ_{j+1}\ldots$:

Definition 10.2 (*number of node occurrences*) Let $u = Q_1Q_2\ldots Q_jQ_{j+1}\ldots \in \boldsymbol{FS}_{\mathbb{X}}$. Define the natural number: $u_j(x) = \sum_{k=1}^{j} \chi_{Q_k^\bullet}(x)$, where $\chi_A(x) = 1$ if $x \in A$ and $\chi_A(x) = 0$ if $x \notin A$ (χ_A is the characteristic function of the set A). It is seen that $u_j(x)$ shows how many times the node x occurred within elements of the prefix $Q_1Q_2\ldots Q_j$. That is, $u_j(x)$ indicates how many times a token enters the node x during execution of $Q_1Q_2\ldots Q_j$.

□

Definition 10.3 (*process, set of processes*) A *process* generated by a firing sequence $u = Q_1Q_2Q_3\ldots \in \boldsymbol{FS}_{\mathbb{X}}$ is defined as $pr(u) = \sum_{j\geq 1} Q_j^{u_j}$ where $Q_j^{u_j}$ is Q_j with each $x \in {}^\bullet Q_j^\bullet$ replaced with the pair $< x, u_j(x) > \in \mathbb{Y} = \mathbb{X} \times \mathbb{N}$. This pair is called an *occurrence* of x in the process $pr(u)$ and $u_j(x)$—a number of x's occurrence in this process. The modified firing component $Q_j^{u_j}$ is an *event*. Let additionally $pr(\varepsilon) = \theta$ and $pr(u) = \Omega_{\mathbb{Y}}$ for $u \notin \boldsymbol{FS}_{\mathbb{X}}$. Processes are denoted by $\alpha, \beta, \gamma, \ldots$ and $\boldsymbol{PR}$ is the set of all processes (subscript $\mathbb{Y}$ is omitted, since processes are always built over $\mathbb{Y}$).

□

Notice, that a process, though motivated by search for a description of c-e structures' behaviour, is defined regardless of any c-e structure whose behaviour it might describe. This allows for collecting processes into sets, i.e. *process languages* and for considering questions like "*does it exist a c-e structure which is a generative device for a given processes language*"—the question analogous to one of the essential questions in formal languages and automata theory. Such topics are considered in Chap. 12. The process definition makes use of the calculus of c-e structures, but processes may also be fully characterised axiomatically [2, 4] as well as inductively. An inductive construction, equivalent to Definition 10.3 but limited to finite processes,

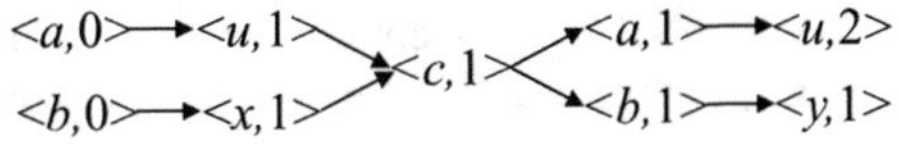

Fig. 10.1 A process evoked by c-e structure $U + V$ in Fig. 8.2

will be provided later in this chapter, its extension to infinite processes—in Chap. 11. Recall (Chap. 6) that "$\sum$" in the process definition, is the least upper bound operator. Directly from the definition of process, we have:

Proposition 10.2 *Any process α with $\theta \neq \alpha \neq \Omega_{\mathbb{Y}}$ is an acyclic, monomial c-e structure over $\mathbb{Y} = \mathbb{X} \times \mathbb{N}$ with contiguous numbering of occurrences of any $x \in \{y | \ \exists k :< y, k >\in car(\alpha)\}$, i.e. if $< x, n >\in car(\alpha)$ and $< x, m >\in car(\alpha)$ and $m > n$ then $< x, k >\in car(\alpha)$ for any k with $n \leq k \leq m$. The numbering starts from 0 or 1. In analogy to Definition 2.2 in Chap. 2:*
$car(\alpha) = \{< z, j >\in \mathbb{Y} : C_\alpha(< z, j >) \neq \theta \vee E_\alpha(< z, j >) \neq \theta\}$

□

For an illustration of a process, let us take the c-e structure $U + V$ in Fig. 8.2 in Chap. 8. Among processes, also infinite, that this structure evokes, is the one in Fig. 10.1. This process (a deadlocked run) is generated by several firing sequences. One of them is, for instance:
$q = Q_1 Q_2 Q_3 Q_4 Q_5 Q_6$ where:

$$Q_1 = (a \to u),\ \ Q_2 = (b \to x),\ \ Q_3 = (u \to c) \bullet (x \to c),$$
$$Q_4 = (c \to a) \bullet (c \to b),\ \ Q_5 = Q_1,\ \ Q_6 = (b \to y)$$

Notice that some permutations of the sequence u generate the same process. Such is, for instance, permutation $Q_2 Q_1 Q_3 Q_4 Q_6 Q_5$. The complete characterization of this phenomenon provides Theorem 10.1. For $v = Q_1 Q_2 \ldots Q_n \in \boldsymbol{FC}^*_{\mathbb{X}}$ denote by $\pi(v)$ a permutation of v, i.e. $\pi(v) = Q_{\pi(1)} Q_{\pi(2)} \cdots Q_{\pi(n)}$ where π is a permutation of sequence $1, 2, \ldots, n$ and denote by $[[v]]$ the composition of relations $[[Q_1]]$, $[[Q_2]], \ldots, [[Q_n]]$: $[[v]] = [[Q_1]] \diamond [[Q_2]] \diamond \ldots \diamond [[Q_n]]$. That is, $(s, t) \in [[v]]$ iff there exist states $s_0, s_1, s_2, \ldots, s_n$ with $s = s_0,\ t = s_n$ and $(s_{j-1}, s_j) \in [[Q_j]]$ for $j = 1, 2, \ldots, n$. Additionally assume $[[\varepsilon]] = id$ (identity relation). Obviously, for any c-e structure U: $[[U]]^* = \bigcup_{u \in FC[U]^*} [[u]]$.

Theorem 10.1 *Let $u \in \boldsymbol{FC}^*_{\mathbb{X}}$, $v = Q_1 Q_2 \ldots Q_n \in \boldsymbol{FC}^*_{\mathbb{X}}$ $(n > 1)$, $w \in \boldsymbol{FC}^*_{\mathbb{X}}$ and let $uvw \in \boldsymbol{FS}_{\mathbb{X}}$. The following conditions (a), (b), (c) are equivalent:*
(a) $^\bullet Q_i^\bullet \cap {}^\bullet Q_j^\bullet = \emptyset$ for $1 \leq i \leq n,\ 1 \leq j \leq n,\ i \neq j$ (that is, Q_i and Q_j are pairwise detached)
(b) $[[uvw]] = [[u\pi(v)w]]$ for any permutation π of v
(c) $pr(uvw) = pr(u\pi(v)w)$ for any permutation π of v.

Proof (a)⇒(b). Let $(s', t') \in [[uvw]] = [[u]] \diamond [[v]] \diamond [[w]]$, thus there are states s, t such that $(s', s) \in [[u]]$, $(s, t) \in [[v]]$, $(t, t') \in [[w]]$. It suffices to show that for any permutation π of v: $(s, t) \in [[\pi(v)]]$, because this would imply $(s', t') \in [[u\pi(v)w]] = [[u]] \diamond [[\pi(v)]] \diamond [[w]]$, thus $[[uvw]] \subseteq [[u\pi(v)w]]$ and, since this holds for each permutation π of v, we get also reverse inclusion, hence $[[uvw]] = [[u\pi(v)w]]$. So, let $(s,t) \in [[v]] = [[Q_1Q_2\dots Q_n]]$. Thus, there exist states $s_0, s_1, s_2, \dots, s_n$ with $s = s_0$, $t = s_n$ and $(s_{j-1}, s_j) \in [[Q_j]]$ for $j = 1, 2, \dots, n$. By Definition 2.7 (of semantics) this yields the following equations:

$s_1 = s_0 \backslash {}^\bullet Q_1 \cup Q_1^\bullet$
$s_2 = s_1 \backslash {}^\bullet Q_2 \cup Q_2^\bullet$
....
$s_n = s_{n-1} \backslash {}^\bullet Q_n \cup Q_n^\bullet$

By substitution we get

$s_2 = (s_0 \backslash {}^\bullet Q_1 \cup Q_1^\bullet) \backslash {}^\bullet Q_2 \cup Q_2^\bullet = (s_0 \backslash {}^\bullet Q_1) \backslash {}^\bullet Q_2 \cup Q_1^\bullet \backslash {}^\bullet Q_2 \cup Q_2^\bullet =$
$s_0 \backslash ({}^\bullet Q_1 \cup {}^\bullet Q_2) \cup (Q_1^\bullet \cup Q_2^\bullet)$

(since the assumption ${}^\bullet Q_i^\bullet \cap {}^\bullet Q_j^\bullet = \emptyset$ implies $Q_1^\bullet \backslash {}^\bullet Q_2 = Q_1^\bullet$). Further substitution similarly yields:

$s_3 = s_0 \backslash ({}^\bullet Q_1 \cup {}^\bullet Q_2 \cup {}^\bullet Q_3) \cup (Q_1^\bullet \cup Q_2^\bullet \cup {}^\bullet Q_3)$ and so on, up to:
....
$s_n = s_0 \backslash ({}^\bullet Q_1 \cup {}^\bullet Q_2 \cup \dots \cup {}^\bullet Q_n) \cup (Q_1^\bullet \cup Q_2^\bullet \cup \dots \cup Q_n^\bullet)$

Since for arbitrary permutation π of $1, 2, \dots, n$:

${}^\bullet Q_1 \cup {}^\bullet Q_2 \cup \dots \cup {}^\bullet Q_n = {}^\bullet Q_{\pi(1)} \cup {}^\bullet Q_{\pi(2)} \cup \dots \cup {}^\bullet Q_{\pi(n)}$ and
$Q_1^\bullet \cup Q_2^\bullet \cup \dots \cup Q_n^\bullet = Q_{\pi(1)}^\bullet \cup Q_{\pi(2)}^\bullet \cup \dots \cup Q_{\pi(n)}^\bullet$, we conclude that
$s_n = s_0 \backslash ({}^\bullet Q_{\pi(1)} \cup {}^\bullet Q_{\pi(2)} \cup \dots \cup {}^\bullet Q_{\pi(n)}) \cup (Q_{\pi(1)}^\bullet \cup Q_{\pi(2)}^\bullet \cup \dots \cup Q_{\pi(n)}^\bullet)$.

Thus, $(s_0, s_n) \in [[Q_{\pi(1)}Q_{\pi(2)}\dots Q_{\pi(n)}]]$, that is $(s,t) \in [[\pi(v)]]$. Therefore $[[uvw]] = [[u\pi(v)w]]$.

(b)⇒(a). Suppose $[[Q_1Q_2\dots Q_n]] = [[Q_{\pi(1)}Q_{\pi(2)}\dots Q_{\pi(n)}]]$ for any permutation π of $1, 2, \dots, n$. This implies $[[Q_iQ_j]] = [[Q_jQ_i]]$ for $i, j = 1, 2, \dots, n$, $i \neq j$ due to $[[Q_1]] \diamond [[Q_2]] \diamond \dots \diamond [[Q_n]] = [[Q_{\pi(1)}]] \diamond [[Q_{\pi(2)}]] \diamond \dots \diamond [[Q_{\pi(n)}]]$ and because each relation $[[Q_j]]$ is a $1-1$ function. Let ${}^\bullet Q_i^\bullet \cap {}^\bullet Q_j^\bullet \neq \emptyset$. We shall infer $[[Q_iQ_j]] \neq [[Q_jQ_i]]$. Let $x \in {}^\bullet Q_i^\bullet \cap {}^\bullet Q_j^\bullet = ({}^\bullet Q_i \cup Q_i^\bullet) \cap ({}^\bullet Q_j \cup Q_j^\bullet) = ({}^\bullet Q_i \cap {}^\bullet Q_j) \cup ({}^\bullet Q_i \cap Q_j^\bullet) \cup (Q_i^\bullet \cap {}^\bullet Q_j) \cup (Q_i^\bullet \cap Q_j^\bullet)$, and let $(s,t) \in [[Q_iQ_j]]$. Obviously, $x \notin {}^\bullet Q_i \cap {}^\bullet Q_j$ and $x \notin Q_i^\bullet \cap Q_j^\bullet$, because Q_iQ_j is a firing sequence (indeed, since assumption $[[Q_1Q_2\dots Q_{n-1}Q_n]] = [[Q_{\pi(1)}Q_{\pi(2)}\dots Q_{\pi(n-1)}Q_{\pi(n)}]]$ implies that $Q_{\pi(1)}Q_{\pi(2)}\dots Q_{\pi(n-1)}Q_{\pi(n)}$ is a firing sequence). Thus, either $x \in {}^\bullet Q_i \cap Q_j^\bullet$ or $x \in Q_i^\bullet \cap {}^\bullet Q_j$. If $x \in {}^\bullet Q_i \cap Q_j^\bullet$ then $(s,t) \notin [[Q_jQ_i]]$, since otherwise $Q_j^\bullet \cap s = \emptyset$ would hold, which is impossible due to $x \in Q_j^\bullet$ and $x \in {}^\bullet Q_i \subseteq s$ (because $(s,t) \in [[Q_iQ_j]]$). If $x \in Q_i^\bullet \cap {}^\bullet Q_j$ then $(s,t) \notin [[Q_jQ_i]]$ since otherwise $x \in s$ would hold (because $x \in {}^\bullet Q_j \subseteq s$), which is impossible due to $Q_i^\bullet \cap s = \emptyset$ (because $(s,t) \in [[Q_iQ_j]]$) and $x \in Q_i^\bullet$. Hence $[[Q_iQ_j]] \neq [[Q_jQ_i]]$, thus we have shown that $[[Q_iQ_j]] = [[Q_jQ_i]] \Rightarrow {}^\bullet Q_i^\bullet \cap {}^\bullet Q_j^\bullet = \emptyset$, hence $[[Q_1Q_2\dots Q_n]] = [[Q_{\pi(1)}Q_{\pi(2)}\dots Q_{\pi(n)}]] \Rightarrow {}^\bullet Q_i^\bullet \cap {}^\bullet Q_j^\bullet = \emptyset$. Since $[[uvw]] = [[u\pi(v)w]] \Leftrightarrow [[v]] = [[\pi(v)]]$, we get ${}^\bullet Q_i^\bullet \cap {}^\bullet Q_j^\bullet = \emptyset$ $i, j = 1, 2, \dots, n$, $i \neq j$.

This ends the verification of equivalence (a)⇔(b).

(a)⇒(c). First, we show (a) $\Rightarrow pr(v) = pr(\pi(v))$. By definition of process: $pr(v) = pr(Q_1 Q_2 \ldots Q_n) = Q_1^{v_1} + Q_2^{v_2} + \cdots + Q_n^{v_n}$ where $v_j = Q_1 Q_2 \ldots Q_j$ $(j = 1, 2, \ldots, n)$ and $Q_j^{v_j}$ is Q_j with each $x \in {}^\bullet Q_j^\bullet$ replaced with $< x, v_j(x) >$ where $v_j(x) = \chi_{Q_1^\bullet}(x) + \chi_{Q_2^\bullet}(x) + \cdots + \chi_{Q_j^\bullet}(x)$ (Definition 10.2). But $\chi_{Q_i^\bullet}(x) = 0$ for $i \neq j$ because, by assumption (a), $x \notin {}^\bullet Q_i^\bullet$, thus $v_j(x) = \chi_{Q_j^\bullet}(x)$. Hence $Q_j^{v_j} = Q_j^{Q_j}$ where $Q_j^{Q_j}$ is Q_j with each $x \in {}^\bullet Q_j^\bullet$ replaced with $< x, \chi_{Q_j^\bullet}(x) >$. But $Q_j^{Q_j} = pr(Q_j)$, thus $pr(v) = \sum_{j=1}^{n} pr(Q_j)$. By commutativity of c-e structures' addition: $\sum_{j=1}^{n} pr(Q_j) = \sum_{j=1}^{n} pr(Q_{\pi(j)})$ for any permutation π of $1, 2, \ldots, n$, hence $pr(v) = pr(\pi(v))$.
Now, it is evident that since the latter equation is invariant of any permutation π, we get $pr(uvw) = pr(u\pi(v)w)$. Indeed, let $u = P_1 P_2 \ldots P_m$ and $w = R_1 R_2 \ldots R_q$. By definition of process: $pr(uvw) = \sum_{j=1}^{m} P_j^{u_j} + \sum_{j=1}^{n} Q_j^{\widehat{v}_j} + \sum_{j=1}^{q} R_j^{\widehat{w}_j}$ where $\widehat{v}_j = P_1 P_2 \ldots P_m Q_1 Q_2 \ldots Q_j$, $\widehat{w}_j = P_1 P_2 \ldots P_m Q_1 Q_2 \ldots Q_n R_1 R_2 \ldots R_j$ and $\widehat{v}_j(x) = u_m(x) + v_j(x)$, $\widehat{w}_j(x) = u_m(x) + v_n(x) + w_j(x)$. But, as shown above, $v_j(x) = \chi_{Q_j^\bullet}(x)$, thus $\widehat{v}_{\pi(j)}(x) = u_m(x) + \chi_{Q_{\pi(j)}^\bullet}(x)$, for any permutation π of $1, 2, \ldots, n$. Hence $\sum_{j=1}^{n} Q_j^{\widehat{v}_j} = \sum_{j=1}^{n} Q_{\pi(j)}^{\widehat{v}_{\pi(j)}}$ where $Q_{\pi(j)}^{\widehat{v}_{\pi(j)}}$ is $Q_{\pi(j)}$ with each $x \in {}^\bullet Q_{\pi(j)}^\bullet$ replaced with $< x, \widehat{v}_{\pi(j)}(x) >$. Therefore, since $pr(u\pi(v)w) = \sum_{j=1}^{m} P_j^{u_j} + \sum_{j=1}^{n} Q_{\pi(j)}^{\widehat{v}_{\pi(j)}} + \sum_{j=1}^{q} R_j^{\widehat{w}_j}$, we get $pr(uvw) = pr(u\pi(v)w)$.

(c)⇒(a). Suppose $pr(Q_1 Q_2 \ldots Q_n) = pr(Q_{\pi(1)} Q_{\pi(2)} \ldots Q_{\pi(n)})$ for any permutation π of $1, 2, \ldots, n$. First, let us show that $pr(Q_i Q_j) = pr(Q_j Q_i)$ for $i, j = 1, 2, \ldots, n$, $i \neq j$. Indeed, the assumption implies

$$pr(Q_1 Q_2 \ldots Q_{n-1} Q_n) = pr(Q_{\pi(1)} Q_{\pi(2)} \ldots Q_{\pi(n-1)} Q_n).$$

The latter equation implies, in turn,

$pr(Q_1 Q_2 \ldots Q_{n-1}) = pr(Q_{\pi(1)} Q_{\pi(2)} \ldots Q_{\pi(n-1)})$, because $pr(Q_1 Q_2 \ldots Q_{n-1} Q_n) = pr(Q_1 Q_2 \ldots Q_{n-1}) + Q_n^{v_n}$ and $pr(Q_{\pi(1)} Q_{\pi(2)} \ldots Q_{\pi(n-1)}) + Q_n^{\widehat{v}_n}$ where $v_n = \chi_{Q_1^\bullet}(x) + \chi_{Q_2^\bullet}(x) + \cdots + \chi_{Q_{n-1}^\bullet}(x) + \chi_{Q_n^\bullet}(x)$ and $\widehat{v}_n = \chi_{Q_{\pi(1)}^\bullet}(x) + \chi_{Q_{\pi(2)}^\bullet}(x) + \cdots + \chi_{Q_{\pi(n-1)}^\bullet}(x) + \chi_{Q_n^\bullet}(x)$, hence $v_n = \widehat{v}_n$. Repeating this reasoning for $n-1, n-2, \ldots, 3, 2$ we get $pr(Q_1 Q_2) = pr(Q_2 Q_1)$ and, since permutation π is arbitrarily chosen, we get $pr(Q_i Q_j) = pr(Q_j Q_i)$ for $i, j = 1, 2, \ldots, n$, $i \neq j$. Second, we show $pr(Q_i Q_j) = pr(Q_j Q_i) \Rightarrow {}^\bullet Q_i^\bullet \cap {}^\bullet Q_j^\bullet = \emptyset$. Assuming ${}^\bullet Q_i^\bullet \cap {}^\bullet Q_j^\bullet \neq \emptyset$ we shall infer $[[Q_i Q_j]] \neq [[Q_j Q_i]]$. Let $x \in {}^\bullet Q_i^\bullet \cap {}^\bullet Q_j^\bullet = ({}^\bullet Q_i \cap {}^\bullet Q_j) \cup ({}^\bullet Q_i \cap Q_j^\bullet) \cup (Q_i^\bullet \cap {}^\bullet Q_j) \cup (Q_i^\bullet \cap Q_j^\bullet)$. Obviously, $x \notin {}^\bullet Q_i \cap {}^\bullet Q_j$ and $x \notin Q_i^\bullet \cap Q_j^\bullet$, because $Q_i Q_j$ is a firing sequence. Thus, either $x \in {}^\bullet Q_i \cap Q_j^\bullet$ or $x \in Q_i^\bullet \cap {}^\bullet Q_j$. For firing components $Q, P \in \mathbf{FC}_{\mathbb{X}}$

such that QP is a firing sequence, denote: ${}^{\circ}pr(QP) = {}^{\bullet}Q \cup {}^{\bullet}P \backslash (Q^{\bullet} \cup P^{\bullet})$ and $pr(QP)^{\circ} = Q^{\bullet} \cup P^{\bullet} \backslash ({}^{\bullet}Q \cup {}^{\bullet}P)$ (for the general case see Proposition 10.3). Obviously, these sets of nodes from $\mathbb{X}$ are defined uniquely. In the case $x \in {}^{\bullet}Q_i \cap Q_j^{\bullet}$, since ${}^{\circ}pr(Q_iQ_j) = {}^{\bullet}Q_i \cup {}^{\bullet}Q_j \backslash (Q_i^{\bullet} \cup Q_j^{\bullet})$ and ${}^{\circ}pr(Q_jQ_i) = {}^{\bullet}Q_j \cup {}^{\bullet}Q_i \backslash (Q_i^{\bullet} \cup Q_j^{\bullet})$, we get $x \in {}^{\circ}pr(Q_iQ_j)$ and $x \notin {}^{\circ}pr(Q_jQ_i)$. In the case $x \in Q_i^{\bullet} \cap {}^{\bullet}Q_j$, since $pr(Q_iQ_j)^{\circ} = Q_i^{\bullet} \cup Q_j^{\bullet} \backslash ({}^{\bullet}Q_i \cup {}^{\bullet}Q_j)$ and $pr(Q_jQ_i)^{\circ} = Q_j^{\bullet} \cup Q_i^{\bullet} \backslash ({}^{\bullet}Q_i \cup {}^{\bullet}Q_j)$, we get $x \in pr(Q_iQ_j)^{\circ}$ and $x \notin pr(Q_jQ_i)^{\circ}$. Therefore, in both cases, $pr(Q_iQ_j) \neq pr(Q_jQ_i)$. Thus, $pr(Q_iQ_j) = pr(Q_jQ_i) \Rightarrow {}^{\bullet}Q_i^{\bullet} \cap {}^{\bullet}Q_j^{\bullet} = \emptyset$. Hence $pr(Q_1Q_2 \ldots Q_n) = pr(Q_{\pi(1)}Q_{\pi(2)} \ldots Q_{\pi(n)}) \Rightarrow {}^{\bullet}Q_i^{\bullet} \cap {}^{\bullet}Q_j^{\bullet} = \emptyset$. Since $pr(uvw) = pr(u\pi(v)w) \Leftrightarrow pr(v) = pr(\pi(v))$, we get ${}^{\bullet}Q_i^{\bullet} \cap {}^{\bullet}Q_j^{\bullet} = \emptyset$ for $i, \ j = 1, 2, \ldots, n, \ i \neq j$.

This ends the verification of equivalence (a)⇔(c) and the whole theorem. □

Remark If we permit the sequence w of firing components in Theorem 10.1 to be infinite ($w \in \boldsymbol{FC}_{\mathbb{X}}^{\omega}$) then point (b) should be rewritten as follows:
(b') $[[uvw_j]] = [[u\pi(v)w_j]]$ for any permutation π of v and any prefix w_j of w.

Let us note that proceesses introduced in Definition 10.3 properly describe behaviour of c-e structures permitting contact situations too, that is such that in a certain state s (reachable from an initial state) a firing component Q may have some nodes in ${}^{\bullet}Q$ and $Q^{\bullet}$ that hold tokens in s. In the following Example, a contact occurs permanently.

Example 10.1 C-e structure in Fig. 10.2 with initial state $s_0 = \{a, c\}$ yields, for instance, the firing sequence in Fig. 10.3, which generates the disconnected process in Fig. 10.4.

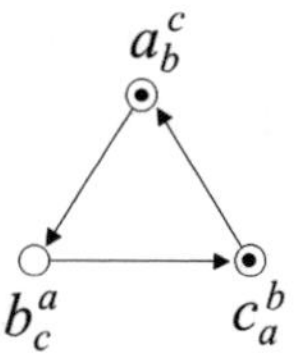

Fig. 10.2 Contact

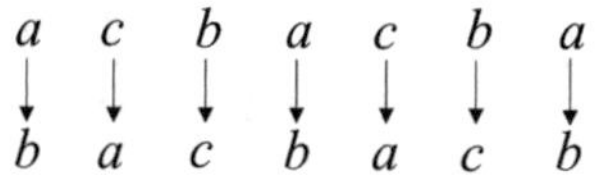

Fig. 10.3 Firing sequence of c-e structure in Fig. 10.2

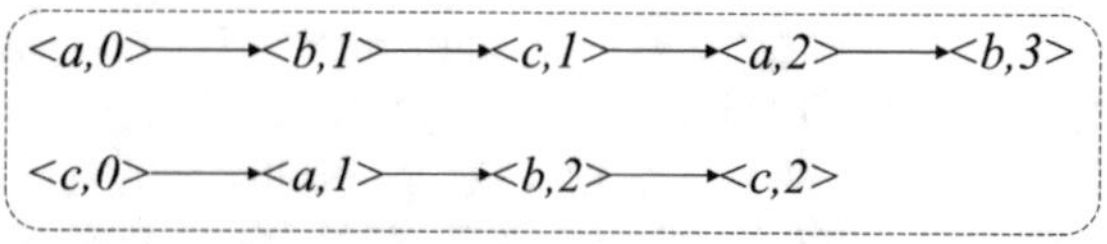

Fig. 10.4 Process generated by firing sequence in Fig. 10.3

Naively, one could define sequential composition of processes—a generalized counterpart of concatenation (juxtaposition) for linear words—by: $\alpha \odot \beta = pr(uv)$ with $\alpha = pr(u)$, $\beta = pr(v)$, $u \in \boldsymbol{FS}_{\mathbb{X}}$, $v \in \boldsymbol{FS}_{\mathbb{X}}$, $uv \in \boldsymbol{FS}_{\mathbb{X}}$ (if $uv \notin \boldsymbol{FS}_{\mathbb{X}}$ then $\alpha \odot \beta$ reduces to $\Omega_{\mathbb{Y}}$). But two problems arise in this case: (1) $\alpha \odot \beta$ would be sensibly defined for finite α only; as we show (Example 11.1) $\alpha \odot \beta$ may be reasonably defined for infinite α too; (2) more importantly, definition $\alpha \odot \beta = pr(uv)$ refers to firing sequences u and v, generating α and β. Processes may be characterised as acyclic c-e structures over $\mathbb{Y} = \mathbb{X} \times \mathbb{N}$ with no choice "+" and with suitable node numbering, as well as inductively, as later in this chapter. Admittance of all such objects as processes requires defining their concatenation in terms of these objects only—without referring to different type of objects, like firing sequences. So, first we formulate two conditions referring to α and β only and equivalent to $uv \in \boldsymbol{FS}_{\mathbb{X}}$—for $|u| < \infty$ (Theorem 10.2). Second, we shall weaken the assumption $|u| < \infty$ (Theorem 10.3), which will allow for defining concatenation $\alpha \odot \beta$ also for infinite α. These two theorems provide conditions ensuring that two firing sequences u and v, when combined, yield a firing sequence. They are essential for a definition of process concatenation, being a total operation on processes. In analogy to the pre and post set of firing component (Definition 2.5), for a process $\alpha = (C_\alpha, E_\alpha) \in \boldsymbol{CE}[\mathbb{Y}]$ with $\mathbb{Y} = \mathbb{X} \times \mathbb{N}$, let us denote:

$$^\circ\alpha^\circ = \{x \mid \exists k :< x, k >\in car(\alpha)\} \qquad \text{(projection of } car(\alpha) \text{ onto } \mathbb{X})$$
$$^\circ\alpha = \{x \in {}^\circ\alpha^\circ \mid \exists k : C_\alpha(< x, k >) = \theta\} \quad \text{(minimal nodes of } \alpha \text{ projected onto } \mathbb{X})$$
$$\alpha^\circ = \{x \in {}^\circ\alpha^\circ \mid \exists k : E_\alpha(< x, k >) = \theta\} \quad \text{(maximal nodes of } \alpha \text{ projected onto } \mathbb{X})$$

For instance, if the process depicted in Fig. 10.1 is α then

$$^\circ\alpha^\circ = \{a, b, c, u, x, y\},$$
$$^\circ\alpha = \{a, b\}$$
$$\alpha^\circ = \{u, y\}$$

The sets $^\circ\alpha$, α° may be also characterised as follows (easy proof is omitted).

Proposition 10.3 *Let $Q_1 Q_2 \ldots \in \boldsymbol{FS}_{\mathbb{X}}$ be a firing sequence, finite or infinite. Then for $\alpha = pr(Q_1 Q_2 \ldots)$:*

$$^\circ\alpha = \bigcup_{j \geq 1} ({}^\bullet Q_j \setminus \bigcup_{i=1}^{j} Q_i^\bullet) = \{x \mid \ < x, 0 >\in car(\alpha)\}$$

$$\alpha^\circ = \bigcup_{j \geq 1} (Q_j^\bullet \setminus \bigcup_{i=1}^{j} {}^\bullet Q_i)$$

□

We may formulate two theorems essential for defining sequential composition of processes, presented in the next chapter.

Theorem 10.2 *Let α and β be processes generated by firing sequences u and v respectively and let $|u| < \infty$. Then uv is a firing sequence if and only if the following conditions hold.*

(i) $^\circ\beta \cap {}^\circ\alpha^\circ \backslash \alpha^\circ = \emptyset$
(ii) $\alpha^\circ \cap {}^\circ\beta^\circ \backslash {}^\circ\beta = \emptyset$
(parentheses in (i) and (ii) are dropped, since $A \cap B \backslash C = (A \cap B)\backslash C$*, for any sets* A, B, C*).*

Proof ($\Rightarrow$). Let $uv \in \boldsymbol{FS}_{\mathbb{X}}$. We prove (i). Let the opposite: $^\circ\beta \cap {}^\circ\alpha^\circ \backslash \alpha^\circ \neq \emptyset$, that is, for a certain x: $x \in {}^\circ\beta$, $x \in {}^\circ\alpha^\circ$, $x \notin \alpha^\circ$. Let $u = Q_1 \dots Q_n$, $v = Q_{n+1}Q_{n+2}, \dots$. $x \in {}^\circ\beta$ implies that there exists $i \in \{n+1, n+2, \dots\}$ such that $x \in {}^\bullet Q_i$. Let k be the *least* of such i's. $x \in {}^\circ\alpha^\circ$ implies that there exists $i \in \{1, \dots, n\}$ such that $x \in {}^\bullet Q_i^\bullet$. Let j be the *greatest* of such i's. $x \notin \alpha^\circ$ and $x \in {}^\bullet Q_j^\bullet$ imply $x \in {}^\bullet Q_j$, since no firing component Q_i $(j < i \leq n)$ contains x. The string $uv = Q_1 \dots Q_j \dots Q_n Q_{n+1} \dots Q_k$. cannot be a firing sequence. Indeed, for $j = n \wedge k = n+1$ this is obvious, for $j < n \wedge k > n+1$, since Q_j is the rightmost member of u containing x and Q_k is the leftmost member of v containing x, we have: $x \in {}^\bullet Q_j$, $x \notin {}^\bullet Q_{j+1}^\bullet, \dots, x \notin {}^\bullet Q_{k-1}^\bullet$, $x \in {}^\bullet Q_k$. Thus, the interval $Q_j \dots Q_k$ of uv is not a firing sequence. Therefore (i) holds. Similarly (ii) is proved.

($\Leftarrow$). Let (i), (ii) hold and let $u = Q_1 \dots Q_n$, $v = P_1 P_2 \dots$. Since $u \in \boldsymbol{FS}_{\mathbb{X}}$, $v \in \boldsymbol{FS}_{\mathbb{X}}$ then each u-computation $s_0', s_1', \dots s_n'$ and each v-computation $t_0', t_1', t_2', \dots$ satisfy $^\circ\alpha \subseteq s_0'$ and $^\circ\beta \subseteq t_0'$. Define $s_0 = {}^\circ\alpha$, $t_0 = {}^\circ\beta$ and consider (a shortest) u-computation:

(u) $s_j = (s_{j-1}\backslash {}^\bullet Q_j) \cup Q_j^\bullet$ for $j = 1, \dots, n$ (thus, $(s_{j-1}, s_j) \in [[Q_j]]$ and $s_n = \alpha^\circ$)
and (a shortest) v-computation:
(v) $t_j = (t_{j-1}\backslash {}^\bullet P_j) \cup P_j^\bullet$ for $j = 1, 2, 3, \dots$ (thus, $(t_{j-1}, t_j) \in [[P_j]]$)
Define:
$s_j' = s_j \cup {}^\circ\beta \backslash {}^\circ\alpha^\circ$ for $j = 0, 1, 2 \dots, n$
$t_j' = t_j \cup \alpha^\circ \backslash {}^\circ\beta^\circ$ for $j = 0, 1, 2, \dots$
To prove $uv \in \boldsymbol{FS}_{\mathbb{X}}$ it suffices to show that:
(1) $s_0', s_1', \dots s_n'$ is a u-computation
(2) $t_0', t_1', t_2', t_3', \dots$ is a v-computation.
(3) $s_n' = t_0'$

Demonstration of (1)
$s_j' = s_j \cup {}^\circ\beta \backslash {}^\circ\alpha^\circ =$ (by (u)) $s_{j-1}\backslash {}^\bullet Q_j \cup Q_j^\bullet \cup {}^\circ\beta\backslash {}^\circ\alpha^\circ = (s_{j-1}\backslash {}^\bullet Q_j \cup {}^\circ\beta\backslash {}^\circ\alpha^\circ) \cup Q_j^\bullet =$ (since $^\circ\beta\backslash {}^\circ\alpha^\circ\backslash {}^\bullet Q_j = {}^\circ\beta\backslash {}^\circ\alpha^\circ$) $= (s_{j-1}\backslash {}^\bullet Q_j \cup {}^\circ\beta\backslash {}^\circ\alpha^\circ\backslash {}^\bullet Q_j) \cup Q_j^\bullet =$ (since $(A \cup B)\backslash C = A\backslash C \cup B\backslash C$) $(s_{j-1} \cup {}^\circ\beta\backslash {}^\circ\alpha^\circ)\backslash {}^\bullet Q_j \cup Q_j^\bullet = s_{j-1}'\backslash {}^\bullet Q_j \cup Q_j^\bullet$. We have also ${}^\bullet Q_j \subseteq s_{j-1}'$ (since ${}^\bullet Q_j \subseteq s_{j-1}$) and $Q_j^\bullet \cap s_{j-1}' = Q_j^\bullet \cap (s_{j-1} \cup {}^\circ\beta^\circ\backslash {}^\circ\alpha^\circ) = (Q_j^\bullet \cap s_{j-1}) \cup (Q_j^\bullet \cap {}^\circ\beta^\circ\backslash {}^\circ\alpha^\circ) =$ (since $Q_j^\bullet \cap s_{j-1} = \emptyset$) $Q_j^\bullet \cap {}^\circ\beta^\circ\backslash {}^\circ\alpha^\circ =$ (since $Q_j^\bullet \subseteq {}^\circ\alpha^\circ$) $\emptyset$. Therefore $(s_{j-1}', s_j') \in [[Q_j]]$ $(j = 1, \dots, n)$ thus (1) holds.

Demonstration of (2)
$t_j' = t_j \cup \alpha^\circ\backslash {}^\circ\beta^\circ =$ (by (v)) $t_{j-1}\backslash {}^\bullet P_j \cup P_j^\bullet \cup \alpha^\circ\backslash {}^\circ\beta^\circ = (t_{j-1}\backslash {}^\bullet P_j \cup \alpha^\circ\backslash {}^\circ\beta^\circ) \cup P_j^\bullet = (t_{j-1}\backslash {}^\bullet P_j \cup \alpha^\circ\backslash {}^\circ\beta^\circ\backslash {}^\bullet P_j) \cup P_j^\bullet = (t_{j-1} \cup \alpha^\circ\backslash {}^\circ\beta^\circ)\backslash {}^\bullet P_j \cup P_j^\bullet = t_{j-1}'\backslash {}^\bullet P_j \cup P_j^\bullet$. We have also ${}^\bullet P_j \subseteq t_{j-1}'$ (since ${}^\bullet P_j \subseteq t_{j-1}$) and $P_j^\bullet \cap t_{j-1}' = P_j^\bullet \cap (t_{j-1} \cup \alpha^\circ\backslash {}^\circ\beta^\circ) =$

$(P_j^\bullet \cap t_{j-1}) \cup (P_j^\bullet \cap \alpha^\circ \backslash {}^\circ\beta^\circ) = P_j^\bullet \cap \alpha^\circ \backslash {}^\circ\beta^\circ = \emptyset$. Therefore $(t'_{j-1}, t'_j) \in [[P_j]]$ $(j = 1, 2, 3, \ldots)$ thus (2) holds.

Demonstration of (3)

Let $x \in s'_n = \alpha^\circ \cup {}^\circ\beta \backslash {}^\circ\alpha^\circ$. If $x \in {}^\circ\beta \backslash {}^\circ\alpha^\circ$ then $x \in {}^\circ\beta$, hence $x \in {}^\circ\beta \cup \alpha^\circ \backslash {}^\circ\beta^\circ = t'_0$.

Let $x \in \alpha^\circ$. The following implications hold: $x \in \alpha^\circ \Rightarrow$ (by (ii)) $\neg x \in {}^\circ\beta^\circ \backslash {}^\circ\beta \Leftrightarrow x \notin {}^\circ\beta^\circ \vee x \in {}^\circ\beta$. If $x \notin {}^\circ\beta^\circ$ then $x \in \alpha^\circ \backslash {}^\circ\beta^\circ$, hence $x \in {}^\circ\beta \cup \alpha^\circ \backslash {}^\circ\beta^\circ = t'_0$. If $x \in {}^\circ\beta$ then $x \in t'_0$. Thus $s'_n \subseteq t'_0$.

Let $x \in t'_0 = {}^\circ\beta \cup \alpha^\circ \backslash {}^\circ\beta^\circ$. If $x \in \alpha^\circ \backslash {}^\circ\beta^\circ$ then $x \in \alpha^\circ$, hence $x \in \alpha^\circ \cup {}^\circ\beta \backslash {}^\circ\alpha^\circ = s'_n$.

Let $x \in {}^\circ\beta$. The following implications hold: $x \in {}^\circ\beta \Rightarrow$ (by (i)) $\neg x \in {}^\circ\alpha^\circ \backslash \alpha^\circ \Leftrightarrow x \notin {}^\circ\alpha^\circ \vee x \in \alpha^\circ$. If $x \notin {}^\circ\alpha^\circ$ then $x \in {}^\circ\beta \backslash {}^\circ\alpha^\circ$, hence $x \in \alpha^\circ \cup {}^\circ\beta \backslash {}^\circ\alpha^\circ = s'_n$. If $x \in \alpha^\circ$ then $x \in t'_n$. Thus $t'_0 \subseteq s'_n$. Therefore $s'_n = t'_0$, that is (3) holds. This ends the proof of the theorem.

□

Assumption $|u| < \infty$ is too restrictive, since we would like to concatenate processes with infinite left constituent too. Thus, we shall weaken this assumption, but for the price, that each firing component occurring in the firing sequence v must have finite pre and post set. To this end, we need an auxiliary notion: $\Psi_\alpha(x)$ which is the number of activations of node x in the process α. Let us associate with each process α a function Ψ_α: $\mathbb{X} \to \mathbb{N}$ defined as:

$$\Psi_\alpha(x) = sup\{k| \quad < x, k > \in car(\alpha)\}$$

Thus, $\Psi_\alpha(x)$ indicates how many times the node x was receiving tokens in the course of process α. Additionally assume: $\Psi_\alpha(x) = 0$ for $x \notin {}^\circ\alpha^\circ$ and $\Psi_\alpha(x) = \infty$ if x occurs infinitely many times in α.

Theorem 10.3 *Let* $u = Q_1 Q_2 \ldots \in \boldsymbol{FS}_{\mathbb{X}}$, $v = P_1 P_2 \ldots \in \boldsymbol{FS}_{\mathbb{X}}$, $\alpha = pr(u)$, $\beta = pr(v)$, $\forall x \in {}^\circ\beta^\circ : \Psi_\alpha(x) < \infty$, $\forall i : |{}^\bullet P_i^\bullet| < \infty$ *and suppose the conditions (i) and (ii) in Theorem 10.2 hold. Then, there exists an interleaving* $w \in \boldsymbol{FS}_{\mathbb{X}}$ *of firing sequences u and v. The converse is false.*

Interleaving of sequences u and v is any sequence w satisfying $w \backslash u = v$ and $w \backslash v = u$, where $w \backslash u$ is what remains of w in effect of deletion of all members of u.

Proof Assumptions $\forall x \in {}^\circ\beta^\circ : \Psi_\alpha(x) < \infty$ and $\forall i : |{}^\bullet P_i^\bullet| < \infty$ imply the existence of a certain number n_1 such that ${}^\bullet Q_j^\bullet \cap {}^\bullet P_1^\bullet = \emptyset$ for $j > n_1$ and the existence of a certain number n_2 such that ${}^\bullet Q_j^\bullet \cap {}^\bullet P_2^\bullet = \emptyset$ for $j > n_2$, etc. One may assume $n_1 < n_2 < n_3 < \ldots$, because if ${}^\bullet Q_j^\bullet \cap {}^\bullet P_k^\bullet = \emptyset$ for $j > n_k$ then ${}^\bullet Q_j^\bullet \cap {}^\bullet P_k^\bullet = \emptyset$ for $j > n_k + m$, for any $m \geq 0$ (the sequence $n_1, n_2, \ldots n_k \ldots$ may be finite or not). Thus, the firing sequence u may be arranged into groups as follows:

$$u = Q_1 \ldots Q_{n_1} \overbrace{Q_{n_1+1} \ldots Q_{n_2} Q_{n_2+1} \ldots Q_{n_k} Q_{n_k+1} Q_{n_k+2} \cdots\cdots}^{\text{no nodes from } P_1 \text{ in this sequence}}$$

no nodes from P_k in this sequence

no nodes from P_2 in this sequence

That is, ${}^{\bullet}Q_j^{\bullet} \cap {}^{\bullet}P_k^{\bullet} = \emptyset$ for $j > n_k$, or, in other words, nodes common to P_k and to u (i.e. to Q_i, $i = 1, 2, \ldots$) may occur in the prefix $Q_1 \ldots Q_{n_k}$ of u only. Indeed, if $x \in {}^{\bullet}P_k^{\bullet}$ occurred arbitrarily far in the sequence $Q_1, Q_2, \ldots$ then, by $\forall i : |{}^{\bullet}P_i^{\bullet}| < \infty$, it would mean that $\Psi_\alpha(x) = \infty$—a contradiction. Define sequence w as:

$$w = \underbrace{Q_1 \ldots Q_{n_1} P_1 Q_{n_1+1} \ldots Q_{n_2} P_2 Q_{n_2+1} \ldots Q_{n_k} P_k}_{w_k} Q_{n_k+1} \ldots Q_{n_{k+1}} \ldots$$

We prove $w \in \boldsymbol{FS}_{\mathbb{X}}$. Let:

$$w_k = Q_1 \ldots Q_{n_1} P_1 Q_{n_1+1} \ldots Q_{n_2} P_2 Q_{n_2+1} \ldots Q_{n_k} P_k \qquad (k > 0)$$

$$w'_k = \underbrace{Q_1 \ldots Q_{n_1} Q_{n_1+1} \ldots Q_{n_2} Q_{n_2+1} \ldots Q_{n_k}}_{\text{prefix of } u} \underbrace{P_1 \ldots P_k}_{\text{prefix of } v}$$

Note that $w_k \in \boldsymbol{FS}_{\mathbb{X}}$ iff $w'_k \in \boldsymbol{FS}_{\mathbb{X}}$. Indeed, sequence w'_k is obtained from w_k by moving all $P_1 \ldots P_k$ past the right end of the sequence $Q_1 \ldots Q_{n_k}$ and, by definition of the sequence w, none of P_j $(j > 0)$, inserted into the sequence u as has been done in w, has an impact on the whole sequence $u = Q_1 Q_2 \ldots$. Certainly, $u_k = Q_1 \ldots Q_{n_k}$ and $v_k = P_1 \ldots P_k$, as prefixes of u and v, are firing sequences. Let $\alpha_k = pr(u_k)$, $\beta_k = pr(v_k)$. Since ${}^{\circ}\beta_k \subseteq {}^{\circ}\beta$, ${}^{\circ}\alpha_k^{\circ}$, we have:
(a) $({}^{\circ}\beta_k \cap {}^{\circ}\alpha_k^{\circ})\backslash\alpha^{\circ} = {}^{\circ}\beta_k\backslash\alpha^{\circ} \cap {}^{\circ}\alpha_k^{\circ}\backslash\alpha^{\circ} \subseteq {}^{\circ}\beta_k \cap {}^{\circ}\alpha_k^{\circ}\backslash\alpha^{\circ} =$ (by (i)) $= \emptyset$. We show:
(b) $({}^{\circ}\beta_k \cap {}^{\circ}\alpha_k^{\circ})\backslash\alpha_k^{\circ} \subseteq ({}^{\circ}\beta_k \cap {}^{\circ}\alpha_k^{\circ})\backslash\alpha^{\circ}$

Let $x \in ({}^{\circ}\beta_k \cap {}^{\circ}\alpha_k^{\circ})\backslash\alpha_k^{\circ}$, that is $x \in {}^{\circ}\beta_k \cap {}^{\circ}\alpha_k^{\circ}$, $x \notin \alpha_k^{\circ}$ and consider the sequence:

$$u = Q_1 \ldots Q_{n_k} \underbrace{Q_{n_k+1} Q_{n_k+2} Q_{n_k+3} \cdots\cdots\cdots}_{\text{no } x \text{ in this sequence because } x \text{ belongs to } {}^{\circ}\beta_k}$$

We have $x \notin {}^{\bullet}Q_j^{\bullet}$ for $j > n_k$, which, along with $x \notin \alpha_k^{\circ}$, implies $x \notin \alpha^{\circ}$. Thus, $x \in ({}^{\circ}\beta_k \cap {}^{\circ}\alpha_k^{\circ})\backslash\alpha_k^{\circ}$ and we have the inclusion (b). By (a) we obtain:
(c) ${}^{\circ}\beta_k \cap {}^{\circ}\alpha_k^{\circ}\backslash\alpha_k^{\circ} = \emptyset$

Similarly, since ${}^{\circ}\beta_k^{\circ} \subseteq {}^{\circ}\beta^{\circ}$, we have:
(d) $(\alpha^{\circ} \cap {}^{\circ}\beta_k^{\circ})\backslash{}^{\circ}\beta = \alpha^{\circ} \cap {}^{\circ}\beta_k^{\circ}\backslash{}^{\circ}\beta \subseteq \alpha^{\circ} \cap {}^{\circ}\beta^{\circ}\backslash{}^{\circ}\beta$ (by (ii)) $= \emptyset$. We show:
(e) $(\alpha_k^{\circ} \cap {}^{\circ}\beta_k^{\circ})\backslash{}^{\circ}\beta_k \subseteq (\alpha^{\circ} \cap {}^{\circ}\beta_k^{\circ})\backslash{}^{\circ}\beta$

Suppose $x \in (\alpha_k^{\circ} \cap {}^{\circ}\beta_k^{\circ})\backslash{}^{\circ}\beta_k$, that is $x \in \alpha_k^{\circ} \cap {}^{\circ}\beta_k^{\circ}$, $x \notin {}^{\circ}\beta_k$ and consider the sequence u as above. Since $x \in {}^{\circ}\beta_k^{\circ}$, we have $x \notin {}^{\bullet}Q_j^{\bullet}$ for $j > n_k$, which, along with $x \in \alpha_k^{\circ}$ implies $x \in \alpha^{\circ}$ and along with $x \notin {}^{\circ}\beta_k$, implies $x \notin {}^{\circ}\beta_k$. Thus, $x \in (\alpha_k^{\circ} \cap {}^{\circ}\beta_k^{\circ})\backslash{}^{\circ}\beta$ and we have inclusion (e). By (d) we obtain:
(f) $\alpha_k^{\circ} \cap {}^{\circ}\beta_k^{\circ}\backslash{}^{\circ}\beta_k = \emptyset$

From (c) and (f), applying Theorem 10.2, we deduce $w'_k \in \boldsymbol{FS}_{\mathbb{X}}$, hence $w_k \in \boldsymbol{FS}_{\mathbb{X}}$. Therefore, each finite prefix of w is a firing sequence, thus, applying Proposition 10.1, we conclude that w is a firing sequence too. A counter example to the converse assertion is:

$u = Q_1 Q_2$ with $Q_1 = c \to a$, $Q_2 = a \to c$, $v = P_1 P_2$ with $P_1 = a \to b$, $P_2 = b \to a$, $w = Q_1 P_1 P_2 Q_2$, $\alpha = pr(u) = < c, 0 > \to < a, 1 > \to < c, 1 >$, $\beta = pr(v) =$
$< a, 0 > \to < b, 1 > \to < a, 1 >$, ${}^\circ\alpha^\circ = \{a, c\}$, $\alpha^\circ = \{c\}$, ${}^\circ\beta = \{a\}$. Thus, $w \in \boldsymbol{FS}_{\mathbb{X}}$ but ${}^\circ\beta \cap {}^\circ\alpha^\circ \backslash \alpha^\circ = \{a\} \neq \emptyset$. This ends the proof of the theorem.

□

For this and the remaining chapters we assume finite pre and post-sets of firing components involved in the considerations. The assumption ensures correctness of the forthcoming construction of concatenation (Definition 11.2 in Chap. 11).

Theorems 10.2 and 10.3 establish conditions (i), (ii), ensuring that two firing sequences, when suitably combined (juxtaposed—Theorem 10.2 and interleaved—Theorem 10.3), yield a firing sequence. The conditions have been formulated exclusively in terms of processes generated by the firing sequences. That is why they will be exploited in definition of process concatenation in the next chapter. Theorem 10.1, on the other hand, implies a property linking any process, i.e. a graph, with an equivalence class of strings that generate this process. In fact, the respective equivalence relation is the one which, in an abstract form, made a starting point to the Mazurkiewicz trace theory [1, 3]:

Define relation $\approx$ between finite firing sequences over $\mathbb{X}$ as:
$u \approx v$ iff $u = u'QPu'' \wedge v = u'PQu'' \wedge {}^\bullet Q^\bullet \cap {}^\bullet P^\bullet = \emptyset$, for some $u' \in \boldsymbol{FC}^*_{\mathbb{X}}$, $u'' \in \boldsymbol{FC}^*_{\mathbb{X}}$, $Q, P \in \boldsymbol{FC}_{\mathbb{X}}$ and let $\overset{*}{\approx}$ (equivalence) be the reflexive and transitive closure of $\approx$. The afore said link is due to:

Theorem 10.4 *For any finite firing sequences u and v:* $u \overset{*}{\approx} v$ *if and only if* $pr(u) = pr(v)$

Proof ($\Rightarrow$). If $u \overset{*}{\approx} v$ then $u = v$ or there are $u_0, u_1, \ldots, u_n$ with $u_0 = u$, $u_n = v$, $u_{j-1} \approx u_j$ for $j = 1, \ldots, n$. But string u_j is obtained by transposition (reversing order) of two consecutive and detached firing components in the string u_{j-1}. Thus, by virtue of Theorem 10.1, we get $u_j \in \boldsymbol{FS}_{\mathbb{X}}$ and $pr(u_{j-1}) = pr(u_j)$ for $j = 1, \ldots, n$, hence $pr(u) = pr(v)$.

($\Leftarrow$). If $pr(u) = pr(v)$ then, by Definition 10.3, strings u and v are composed of the same firing components, lined up, possibly in a different order in u and in v. Thus $v = \pi(u)$ for a certain permutation π. For $u = v$ the theorem obviously holds, for $u \neq v$ each permutation is always a composition of transpositions of consecutive elements. Let us take into account transpositions leading from u to v, which preserve equality of processes generated by respective strings (such transpositions must exist, since $pr(u) = pr(v)$). By Theorem 10.1, transposed are detached firing components only. But, if two strings differ by two consecutive detached firing components only, then relation $\approx$ between them holds. This implies $u \overset{*}{\approx} v$.

□

The reader familiar with Mazurkiewicz trace will immediately recognise that the $\overset{*}{\approx}$-equivalence classes of finite firing sequences composed of firing components of a given c-e structure U, are Mazurkiewicz traces over a concurrent alphabet $(\boldsymbol{FC}_{\mathbb{X}}[U], I)$ with independence relation $I \subseteq \boldsymbol{FC}_{\mathbb{X}} \times \boldsymbol{FC}_{\mathbb{X}}$ defined by $(Q, P) \in I$ iff $^\bullet Q^\bullet \cap {}^\bullet P^\bullet = \emptyset$. Therefore, by virtue of Theorem 10.4, there is a 1-1 correspondence between finite processes and Mazurkiewicz traces represented by firing sequences generated by U.

Now, let us notice that processes immediately determine the relation of causal dependence and its opposite—concurrency. These relations cannot be just adopted from their counterparts in Petri net theory, because our processes are also disconnected graphs, with causal dependencies possible also between events belonging to disconnected parts, e.g in case of systems admitting contact situations (but not only!). As stated in Proposition 10.2, any process α with $\theta \neq \alpha \neq \Omega_{\mathbb{Y}}$ is an acyclic c-e structure over $\mathbb{Y} = \mathbb{X} \times \mathbb{N}$. Thus, the set of firing components in α is $\boldsymbol{FC}_{\mathbb{Y}}[\alpha]$. These firing components are *events*. The causal dependence and concurrency are established between events:

Definition 10.4 (*causal dependence and concurrency*) Let a process α be given. For an event $q = (C_q, E_q) \in \boldsymbol{FC}_{\mathbb{Y}}[\alpha]$ denote:
$^\bullet q = \{< x, k > \in car(q) |\ C_q(< x, k >) = \theta\}$
$q^\bullet = \{< x, k > \in car(q) |\ E_q(< x, k >) = \theta\}$
Let $q, p \in \boldsymbol{FC}_{\mathbb{Y}}[\alpha]$ be events in the process α. We say that q *causally precedes* (or equals) p in α, written $q \preceq_\alpha p$, iff $q = p$ or there are events $q_0, q_1, \ldots, q_n$ $(n > 0)$ in α satisfying $q = q_0,\ p = q_n$ and
$\forall j \in \{1, \ldots, n\} : [q^\bullet_{j-1} \cap {}^\bullet q_j \neq \emptyset \vee \exists x, k : (< x, k > \in car(q_{j-1}) \wedge < x, k+1 > \in car(q_j))]$.

Events q and p in α are *causally dependent* iff either $q \preceq_\alpha p$ or $p \preceq_\alpha q$ and *concurrent* otherwise.

□

Note that the right operand of the disjunction "$\vee$" in this definition is introduced for the case of disconnected processes—see Example 10.1. It is easy to verify that causal precedence is a partial order in a process: $q \preceq_\alpha q$ and $(q \preceq_\alpha p \wedge p \preceq_\alpha r) \Rightarrow q \preceq_\alpha r$ are obvious, $(q \preceq_\alpha p \wedge p \preceq_\alpha q) \Rightarrow q = p$ follows from the cycle-freeness of graphs-processes and from increasing numbering of node occurrences.

Concluding this chapter, let us give another process definition, the inductive one. Here, we formulate it for finite processes, then in Chap. 11 it will be extended to infinite processes.

Definition 10.5 (*atomic process*) Let $Q \in \boldsymbol{FC}_{\mathbb{X}}$ be a firing component over $\mathbb{X}$. An *atomic process* q, generated by Q, is Q with each $x \in {}^\bullet Q$ replaced with the pair $< x, 0 >$ and each $x \in Q^\bullet$ replaced with the pair $< x, 1 >$. Denote: $^\circ q = {}^\bullet Q$, $q^\circ = Q^\bullet$, $^\circ q^\circ = {}^\circ q \cup q^\circ$, $\mathbb{Y}_q = {}^\circ q \times \{0\} \cup q^\circ \times \{1\}$ and define function $\Psi_q : \mathbb{X} \rightarrow \{0, 1\}$ by:

$$\Psi_q(x) = \begin{cases} 1 \text{ if } x \in q^\circ \\ 0 \text{ else} \end{cases}$$

The set of all atomic processes is denoted by $\boldsymbol{PR}_0$.

□

Notice that Q is a c-e structure over $\mathbb{X}$, thus the atomic process q is a c-e structure over $\mathbb{Y} = \mathbb{X} \times \mathbb{N}$.

Definition 10.6 (*inductive construction of finite process*) 1. Atomic process, the neutral θ and the chaos $\Omega_{\mathbb{Y}}$ (see the note following Proposition 10.1) is a finite process. Assume $^{\circ}\theta^{\circ} = {}^{\circ}\theta = \theta^{\circ} = \mathbb{Y}_{\theta} = \emptyset$, $\Psi_{\theta}(x) = 0$ for each $x \in \mathbb{X}$ and $^{\circ}\Omega_{\mathbb{Y}}^{\circ} = {}^{\circ}\Omega_{\mathbb{Y}} = \Omega_{\mathbb{Y}}^{\circ} = \mathbb{X}$, $\mathbb{Y}_{\Omega_{\mathbb{Y}}} = \mathbb{Y} = \mathbb{X} \times \mathbb{N}$.
2. Let α be a finite process already defined (a c-e structure over $\mathbb{Y} = \mathbb{X} \times \mathbb{N}$.) and q an atomic process. Thus, sets $^{\circ}\alpha^{\circ}$, $^{\circ}\alpha$, α°, $^{\circ}q^{\circ}$, $^{\circ}q$, q°, $\mathbb{Y}_{\alpha}$ and, for $\alpha \neq \Omega_{\mathbb{Y}}$ the function Ψ_{α}, are defined. Then, the q-successor of α defined as:

$$suc_q(\alpha) = \begin{cases} \alpha + q^{\alpha} & \text{if } \alpha^{\circ} \cap {}^{\circ}q^{\circ} \backslash {}^{\circ}q = \emptyset = {}^{\circ}q \cap {}^{\circ}\alpha^{\circ} \backslash \alpha^{\circ} \\ \Omega_{\mathbb{Y}} & \text{else} \end{cases}$$

is a process, where for $\alpha \neq \Omega_{\mathbb{Y}}$, by q^{α} is denoted q in which each pair $< x, k > \in \mathbb{Y}_q$ ($k \in \{0, 1\}$) is replaced with $< x, k + \Psi_{\alpha}(x) >$. Obviously, $suc_q(\Omega_{\mathbb{Y}}) = \Omega_{\mathbb{Y}}$, $suc_q(\theta) = q$. Notice that because α and q^{α} are c-e structures (over $\mathbb{Y}$), the summation (+) in Definition of $suc_q(\alpha)$ makes sense. Define:

$\alpha \odot q = suc_q(\alpha)$
$^{\circ}suc_q(\alpha)^{\circ} = {}^{\circ}\alpha^{\circ} \cup {}^{\circ}q^{\circ}$
$^{\circ}suc_q(\alpha) = {}^{\circ}\alpha \cup ({}^{\circ}q \backslash \alpha^{\circ})$
$suc_q(\alpha)^{\circ} = q^{\circ} \cup (\alpha^{\circ} \backslash {}^{\circ}q)$ and for $\alpha \odot q \neq \Omega_{\mathbb{Y}}$:
$\mathbb{Y}_{\alpha \odot q} = \mathbb{Y}_{\alpha} \cup \mathbb{Y}_{q^{\alpha}}$ where $\mathbb{Y}_{q^{\alpha}}$ is $\mathbb{Y}_q$ with each $< x, k > \in \mathbb{Y}_q$ ($k \in \{0, 1\}$) replaced with $< x, k + \Psi_{\alpha}(x) >$
$\Psi_{\alpha \odot q}(x) = \Psi_{\alpha}(x) + \Psi_q(x)$

The set of all finite processes is denoted by $\boldsymbol{PR}_{fin}$.

□

Notice that $\Psi_{\alpha}(x) = sup\{k| < x, k > \in \mathbb{Y}_{\alpha}\}$ for any $\alpha \in PR_{fin}$ and $x \in \mathbb{X}$, where $\mathbb{Y}_{\alpha}$, as defined above, is the set of all pairs $< x, k >$ in the process α. Number $\Psi_{\alpha}(x)$ is an increment, or "shift", by which pairs $< x, n >$ ($n \in \{0, 1\}$) occuring in the atomic process q must be modified to obtain process $\alpha \odot q$. Obviously, Ψ_{α} here defined inductively, coincides with the function (identically named) used in Theorem 10.3.

The reader is encouraged to verify that each finite process α (i.e. such that $|car(\alpha)| < \infty$) constructed in Definition 10.3, is a process constructed in Definition 10.6 and conversely.

References

1. Diekert V, Rozenberg G (eds) (1995) The book of traces. World Scientific, Singapore, New Jersey, London, Hong Kong
2. Kusmirek A (1996) Axiomatic characterisation and some properties of processes in cause-effect nets (Levi's lemma, pumping lemma and others), M.Sc. thesis (in Polish), Warsaw University
3. Mazurkiewicz A (1987) Trace theory. In Brauer W et al (eds) Petri nets. Applications and relationship to other models of concurrency, number 255, in Lecture Notes in Computer Science. Springer, Berlin, Heidelberg, New York, pp 279–324
4. Raczunas M (2000) Algebra procesów generowanych przez struktury przyczyn i skutków, (*Algebra of processes generated by cause-effect structures)*, submitted as a PhD thesis (in Polish), Warsaw University

Chapter 11
Monoid of Processes

A monoid of processes is the algebra $<\boldsymbol{PR}, \odot, \theta>$ where $\boldsymbol{PR}$ is the set of all processes being a special form of c-e structures defined in Chap. 10, θ is the neutral element for addition and multiplication of c-e structures (Proposition 2.1, Chap. 2) and $\odot$ is a concatenation of processes, with θ being a neutral for $\odot$ too. This chapter comprises definition and properties of process concatenation. Concatenation $\alpha \odot \beta$ denotes a sequential composition, meant as juxtaposition of processes α and β with "gluing up" some effects of maximal (w.r.t. $\leq_\alpha$—Definition 10.4, Chap. 10) events in α with identical causes of minimal (w.r.t. $\leq_\beta$) events in β. Not always, however, such composition yields a meaningful process, that is, such one for which a generative firing sequence exists. Nonetheless, it will be convenient to have concatenation as a total operator, and even is necessary for the subject of the next chapter. Moreover, for infinite α's, the juxtaposition would require introducing "transfinite" processes, a concept of lower intuitive appeal than that of "real" processes presenting records of c-e structures' activities. Sometimes, as Example 11.1 shows, combining one-by-one infinite processes yields intuitively acceptable result. We aim at a rigorous concept of concatenation, simple and concise, but capturing all these intuitive requirements. To this end, conditions (i), (ii) from Theorem 10.2 will be essentially exploited. Recall that $\Omega_{\mathbb{Y}}$ symbolises "chaos", i.e. an "unstructured" c-e structure satisfying $V + \Omega_{\mathbb{Y}} = \Omega_{\mathbb{Y}}$ for any c-e structure V over $\mathbb{Y} = \mathbb{X} \times \mathbb{N}$ and define an auxiliary notion:

Definition 11.1 (β^α: "*shift* β *by* α") Let α, β be c-e structures over $\mathbb{Y} = \mathbb{X} \times \mathbb{N}$, processes in particular, with $\forall x \in {}^\circ\beta^\circ : \Psi_\alpha(x) < \infty$ where ${}^\circ\beta^\circ = \{x \mid \exists k :< x, k > \in car(\beta)\}$ is the projection of β onto $\mathbb{X}$, $\Psi_\alpha(x) = sup\{k \mid \ < x, k > \in car(\alpha)\}$. By β^α is denoted β with each node $< x, k > \in car(\beta)$ replaced with $< x, k + \Psi_\alpha(x) >$. Additionally, assume $\beta^\alpha = \Omega_{\mathbb{Y}}$ if $\beta = \Omega_{\mathbb{Y}} \vee \exists x \in {}^\circ\beta^\circ : \Psi_\alpha(x) = \infty$ and $\theta^\alpha = \theta$ for $\alpha \neq \Omega_{\mathbb{Y}}$.

L. Czaja, *Cause-Effect Structures*, Lecture Notes in Networks and Systems 45, https://doi.org/10.1007/978-3-030-20461-7_11

Definition 11.2 (*concatenation of processes*) For processes $\alpha, \beta \in \boldsymbol{CE}[\mathbb{Y}]$ concatenation $\alpha \odot \beta$ is defined as:

$$\alpha \odot \beta = \begin{cases} \alpha + \beta^{\alpha} \text{ if } \alpha^{\circ} \cap {}^{\circ}\beta^{\circ} \backslash {}^{\circ}\beta = \emptyset = {}^{\circ}\beta \cap {}^{\circ}\alpha^{\circ} \backslash \alpha^{\circ} \\ \Omega_{\mathbb{Y}} \quad \text{else} \end{cases}$$

On account of the forthcoming Theorem 11.2, we shall further denote $\Omega_{\mathbb{Y}}$ by $\mathbb{O}$. □

Before showing that the concatenation is well defined, that is, it yields a process distinct from $\Omega_{\mathbb{Y}}$, when α and β are not $\Omega_{\mathbb{Y}}$ and meet conditions (i), (ii) from Theorem 10.2 and $\forall x \in {}^{\circ}\beta^{\circ} : \Psi_{\alpha}(x) < \infty$, let us illustrate this by the following Example 11.1:

11.1 Example

Example 11.1 shows that $\alpha \odot \beta \neq \Omega_{\mathbb{Y}}$ may happen for infinite α. Note that $\beta \odot \alpha = \Omega_{\mathbb{Y}}$, because $\Psi_{\beta}(c) = \infty$ (Fig. 11.1).

Theorem 11.1 *If α and β are processes then $\alpha \odot \beta$ is a process. If $\alpha \odot \beta \neq \Omega_{\mathbb{Y}}$ then $\alpha \odot \beta$ is generated by a firing sequence w, whose existence has been stated in Theorem 10.3.*

Proof If $\alpha = \Omega_{\mathbb{Y}} \vee \beta = \Omega_{\mathbb{Y}} \vee {}^{\circ}\beta \cap {}^{\circ}\alpha^{\circ} \backslash \alpha^{\circ} \neq \emptyset \vee \alpha^{\circ} \cap^{\circ} \beta^{\circ} \backslash {}^{\circ}\beta \neq \emptyset \vee \neg\forall x \in^{\circ} \beta^{\circ} : \Psi_{\alpha}(x) < \infty$ then $\alpha \odot \beta = \Omega_{\mathbb{Y}} \in \boldsymbol{PR}$. Suppose $u = Q_1 Q_2 .... \in \boldsymbol{FS}_{\mathbb{X}}$, $v = P_1 P_2 .... \in \boldsymbol{FS}_{\mathbb{X}}$, $\alpha = pr(u)$, $\beta = pr(v)$, ${}^{\circ}\beta \cap {}^{\circ}\alpha^{\circ} \backslash \alpha^{\circ} = \emptyset$, $\alpha^{\circ} \cap {}^{\circ}\beta^{\circ} \backslash {}^{\circ}\beta = \emptyset$, $\forall x \in {}^{\circ}\beta^{\circ} : \Psi_{\alpha}(x) < \infty$, $\gamma = \alpha \odot \beta = \alpha + \beta^{\alpha}$. Recall that we have assumed finite pre and post-sets of each firing component involved in the considerations.

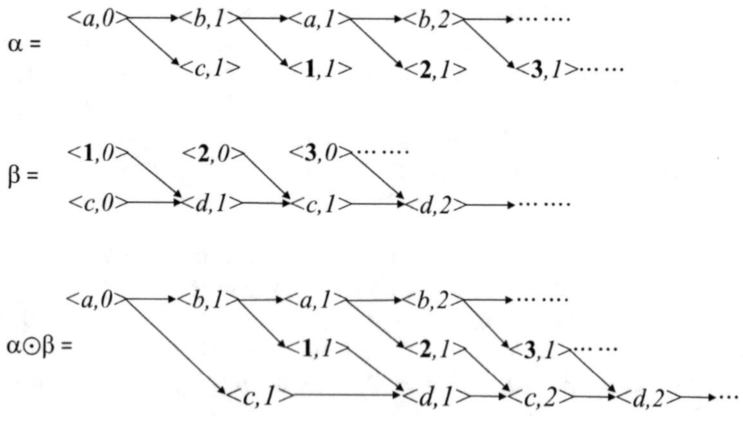

Fig. 11.1 Concatenation of infinite processes. Symbols a, b, c, d and bold numerals **1**, **2**, **3**, ... are elements of the set $\mathbb{X}$; Occurrence number $k = 1$ for nodes $< \mathbf{1,k} >, < \mathbf{2,k} >, < \mathbf{3,k} >$, ... in processes α and $\alpha \odot \beta$, and $k = 0$ in β; Occurrence number $k = 0, 1, 2, 3,$ ..... for nodes $< a, k >$ in α and $\alpha \odot \beta$, and for nodes $< c, k >$ in β; Occurrence number $k = 1, 2, 3,$ ..... for nodes $< c, k >$ in processes α and $\alpha \odot \beta$ and for nodes $< d, k >$ in β and $\alpha \odot \beta$. Therefore $\Psi_{\alpha}(a) = \Psi_{\alpha}(b) = \infty$, $\Psi_{\alpha}(c) = \Psi_{\alpha}(\mathbf{1}) = {}_{\alpha}(\mathbf{2}) = {}_{\alpha}(\mathbf{3}) = ... = \mathbf{1}$, $\Psi_{\alpha}(d) = 0$

For demonstration of $\gamma \in \boldsymbol{PR}$, we have to find a firing sequence w with $\gamma = pr(w)$. We claim that w constructed in the proof of Theorem 10.3, i.e.

$$w = \underbrace{Q_1....Q_{n_1} P_1 Q_{n_1+1}....Q_{n_2} P_2 Q_{n_2+1}....Q_{n_k} P_k}_{w_k} Q_{n_k+1}....Q_{n_{k+1}}....$$

is such sequence (for the meaning of numbers n_k refer to the proof of Theorem 10.3). On account of this Theorem, $w \in \boldsymbol{FS}_{\mathbb{X}}$ thus, by Definition 10.3 (of process) and by $^{\bullet}Q_j^{\bullet} \cap {}^{\bullet}P_k^{\bullet} = \emptyset$ for $j > n_k$ $(k = 1, 2, ...)$:

$$pr(w) = \sum_{j=1}^{n_1} Q_j^{u_j} + P_1^{w_1} + \sum_{j=n_1+1}^{n_1} Q_j^{u_j} + P_2^{w_2} + ... + \sum_{j=n_{k-1}+1}^{n_k} Q_j^{u_j} + P_k^{w_k} + ...$$

where $u_j = Q_1 Q_2....Q_j$, $w_k = Q_1....Q_{n_1} P_1 Q_{n_1+1}....Q_{n_2} P_2 Q_{n_2+1}....Q_{n_k} P_k$, Thus,

$$pr(w) = \sum_{j=1}^{|u|} Q_j^{u_j} + \sum_{k=1}^{|v|} P_k^{w_k}$$

By definition of α and β: $\alpha = pr(u) = \sum_{j=1}^{|u|} Q_j^{u_j}$ and $\beta = \sum_{k=1}^{|v|} P_k^{v_k}$ where $v_k = P_1 P_2....P_k$.

By Definition 10.3, $P_k^{w_k}$ is P_k with each $x \in car(P_k)$ replaced with $< x, w_k(x) >$. But by $^{\bullet}Q_j^{\bullet} \cap {}^{\bullet}P_k^{\bullet} = \emptyset$ for $j > n_k$ $(k = 1, 2, ...)$ and definition of $w_k(x)$: $w_k(x) = v_k(x) + \Psi_\alpha(x)$. Hence $P_k^{w_k}$ is P_k with each $x \in car(P_k)$ replaced with $< x, v_k(x) + \Psi_\alpha(x) >$, which implies, by definition of β^α:

$\beta^\alpha = \sum_{k=1}^{|v|} P_k^{w_k}$. Finally, $\gamma = \alpha \odot \beta = \alpha + \beta^\alpha = pr(w) \in \boldsymbol{PR}$. □

For proving associativity of concatenation in Theorem 11.2, we need the following three propositions:

Proposition 11.1 *Let* $\alpha = (C_\alpha, E_\alpha)$ *be a process and let* $< x, k > \in car(\alpha)$, $< x, k' > \in car(\alpha)$. *Then:*

(a) $C_\alpha(< x, k >) = \theta = C_\alpha(< x, k' >) \Rightarrow k = k'$
(b) $E_\alpha(< x, k >) = \theta = E_\alpha(< x, k' >) \Rightarrow k = k'$

That is, among the causes of minimal (w.r.t. $\leq_\alpha$*—Definition 10.4) events in* α *there is at most one node* $< x, k >$ *with a given* x*; the same holds for effects of maximal events in* α*.*

Proof **Of (a)**—obvious, since $C_\alpha(< x, k >) = \theta \Leftrightarrow k = 0$.

Of (b) Let $\alpha = pr(u)$ with $u = Q_1 Q_2.... \in \boldsymbol{FS}_{\mathbb{X}}$, and let $u_i = Q_1 Q_2....Q_i$ $(i > 0)$. By Definition 10.3 (of process):
$(< x, k > \in car(\alpha) \wedge E_\alpha(< x, k >) = \theta) \Rightarrow \exists! j : (k = u_j(x) \wedge x \in Q_j^{\bullet} \wedge \forall i > j : x \notin {}^{\bullet}Q_i^{\bullet})$
and
$(< x, k' > \in car(\alpha) \wedge E_\alpha(< x, k' >) = \theta) \Rightarrow \exists! j' : (k' = u_{j'}(x) \wedge x \in Q_{j'}^{\bullet} \wedge \forall i > j' : x \notin {}^{\bullet}Q_i^{\bullet})$
where "∃!" means "there exists unique". Assume $E_\alpha(< x, k >) = \theta$ and $E_\alpha(< x, k' >) = \theta$ and let j and j' be numbers whose existence has been

asserted under these assumptions. Suppose $k \neq k'$. Then (by definition of numbers j and j') $j \neq j'$. But $j < j'$ contradicts $x \in Q_j^\bullet \wedge \forall i > j : x \notin {}^\bullet Q_i^\bullet$ and $j' < j$ contradicts $x \in Q_{j'}^\bullet \wedge \forall i > j' : x \notin {}^\bullet Q_i^\bullet$. Hence $j = j'$, which implies $k = k'$. □

Proposition 11.2 *For c-e structures* $\alpha, \beta, \gamma \in \boldsymbol{CE}[\mathbb{Y}]$ *with* $\mathbb{Y} = \mathbb{X} \times \mathbb{N}$ *(processes in particular):*

(a) $(\beta + \gamma)^\alpha = \beta^\alpha + \gamma^\alpha$
(b) $\Psi_{\alpha+\beta}(x) = max(\Psi_\alpha(x), \Psi_\beta(x))$ *for each* $x \in \mathbb{X}$
(c) $\Psi_{\beta^\alpha}(x) = \Psi_\beta(x) + \Psi_\alpha(x)$ *for each* $x \in \mathbb{X}$

Proof **Of (a)**. It may be assumed that $\Psi_\alpha(x) < \infty$ for $x \in {}^\circ\beta^\circ$ and for $x \in {}^\circ\gamma^\circ$, since otherwise both sides of equation (a) are $\Omega_\mathbb{Y}$. By Definition 11.1, β^α is β with each $< x, k > \in car(\beta)$ replaced with $< x, k + \Psi_\alpha(x) >$ and γ^α is γ with each $< x, k > \in car(\gamma)$ replaced with $< x, k+\Psi_\alpha(x) >$. Hence, $\beta^\alpha+\gamma^\alpha$ is $\beta+\gamma$ with each $< x, k > \in car(\beta) \cup car(\gamma)$ replaced with $< x, k + \Psi_\alpha(x) >$. Since $car(\beta) \cup car(\gamma) = car(\beta + \gamma)$, then by Definition 11.1 we obtain (a).
Of (b). $\Psi_{\alpha+\beta}(x) = sup\{k \mid < x, k > \in car(\alpha + \beta)\} = sup\{k \mid < x, k > \in car(\alpha) \cup car(\beta)\} = max(sup\{k \mid < x, k > \in car(\alpha)\}, sup\{k \mid < x, k > \in car(\beta)\}) = max(\Psi_\alpha(x), \Psi_\beta(x))$.
Of (c). $\Psi_{\beta^\alpha}(x) = sup\{k \mid < x, k > \in car(\beta^\alpha)\} =$ (by Definition 11.1) $sup\{j + \Psi_\alpha(x) \mid < x, j > \in car(\beta)\} = sup\{j \mid < x, j > \in car(\beta)\} + \Psi_\alpha(x)$ $= \Psi_\beta(x) + \Psi_\alpha(x)$. □

Proposition 11.3 *For any processes* α, β, γ*:* $\alpha \odot (\beta \odot \gamma) = \Omega_\mathbb{Y} \Leftrightarrow (\alpha \odot \beta) \odot \gamma = \Omega_\mathbb{Y}$

Proof For $\alpha = \Omega_\mathbb{Y} \vee \beta = \Omega_\mathbb{Y} \vee \beta \odot \gamma = \Omega_\mathbb{Y} \vee \alpha \odot \beta = \Omega_\mathbb{Y}$ the proposition holds evidently. Suppose then $\alpha \neq \Omega_\mathbb{Y} \wedge \beta \neq \Omega_\mathbb{Y} \wedge \gamma \neq \Omega_\mathbb{Y} \wedge \beta \odot \gamma \neq \Omega_\mathbb{Y} \wedge \alpha \odot \beta \neq \Omega_\mathbb{Y}$. If $\Psi_\alpha(x)=\infty$ for a certain $x \in {}^\circ(\beta\odot\gamma)^\circ$ then $\alpha \odot (\beta \odot \gamma) = \Omega_\mathbb{Y}$. In this case, since ${}^\circ(\beta\odot\gamma)^\circ = {}^\circ\beta^\circ \cup {}^\circ\gamma^\circ$, we have $x \in {}^\circ\beta^\circ \vee x \in {}^\circ\gamma^\circ$, but $x \notin {}^\circ\beta^\circ$ (because $\Psi_\alpha(x) = \infty$ and $\alpha \odot \beta \neq \Omega_\mathbb{Y}$) hence $x \in {}^\circ\gamma^\circ$. Since $x \in {}^\circ\alpha^\circ$, we conclude $\Psi_{\alpha\odot\beta}(x) = \infty$, which implies $(\alpha \odot \beta) \odot \gamma = \Omega_\mathbb{Y}$. Similarly we check that: if $\Psi_{\alpha\odot\beta}(x) = \infty$ for a certain $x \in {}^\circ\gamma^\circ$ then $(\alpha \odot \beta) \odot \gamma = \Omega_\mathbb{Y}$ holds and this implies $\alpha \odot (\beta \odot \gamma) = \Omega_\mathbb{Y}$. Suppose, then, also that $\forall x \in {}^\circ(\beta \odot \gamma)^\circ : \Psi_\alpha(x) < \infty$ and $\forall x \in {}^\circ\gamma^\circ : \Psi_{\alpha\odot\beta}(x) < \infty$. By Definition 11.2 (of process concatenation), what remains to be shown is: (a)$\wedge$(b) $\Leftrightarrow$ (c)$\wedge$(d) where:

(a) ${}^\circ(\beta \odot \gamma) \cap {}^\circ\alpha^\circ \backslash \alpha^\circ = \emptyset$ (c) $\gamma^\circ \cap {}^\circ(\alpha \odot \beta)^\circ \backslash (\alpha \odot \beta)^\circ = \emptyset$
(b) $\alpha^\circ \cap {}^\circ(\beta \odot \gamma)^\circ \backslash {}^\circ(\beta \odot \gamma) = \emptyset$ (d) $(\alpha \odot \beta)^\circ \cap {}^\circ\gamma^\circ \backslash {}^\circ\gamma = \emptyset$

Equivalently, we show that (a')$\vee$(b') $\Leftrightarrow$ (c')$\vee$(d') where:

(a') $\exists x : [x \in {}^\circ(\beta \odot \gamma) \wedge x \in {}^\circ\alpha^\circ \backslash \alpha^\circ]$ (c') $\exists x : [x \in \gamma^\circ \wedge x \in {}^\circ(\alpha \odot \beta)^\circ \backslash (\alpha \odot \beta)^\circ]$
(b') $\exists x : [x \in \alpha^\circ \wedge {}^\circ(\beta \odot \gamma)^\circ \backslash {}^\circ(\beta \odot \gamma)]$ (d') $\exists x : [x \in (\alpha \odot \beta)^\circ \wedge x \in {}^\circ\gamma^\circ \backslash {}^\circ\gamma]$

Let (a') hold. From $x \in {}^\circ(\beta \odot \gamma) \wedge x \in {}^\circ\alpha^\circ \backslash \alpha^\circ$ we infer $x \in {}^\circ\gamma$, since $x \notin {}^\circ\beta$. The latter holds indeed, since $x \in {}^\circ\beta$ would imply $x \in {}^\circ\beta \cap {}^\circ\alpha^\circ \backslash \alpha^\circ$, hence $^\circ\beta \cap {^\circ\alpha^\circ} \backslash \alpha^\circ \notin \emptyset$ and by Definition 11.2 we would have $\alpha \odot \beta = \Omega_{\mathbb{Y}}$. But $x \notin \beta^\circ$ as well, because if $x \in \beta^\circ$ then, by $x \in {^\circ\gamma} \wedge x \notin {^\circ\beta}$ and by Proposition 11.1, we would arrive to $x \notin {^\circ(\beta \odot \gamma)}$—a contradiction. From $x \in {^\circ\alpha^\circ} \backslash \alpha^\circ$ and $x \notin \beta^\circ$ we infer $x \in (\alpha \odot \beta)^\circ$, hence $x \in {^\circ(\alpha \odot \beta)^\circ} \backslash (\alpha \odot \beta)^\circ$, implying that (c') holds.

Let (b') hold. $x \in \alpha^\circ \wedge x \in {^\circ(\beta \odot \gamma)^\circ} \backslash {^\circ(\beta \odot \gamma)}$ implies $x \notin {^\circ\beta^\circ}$. Indeed, if $x \in {^\circ\beta^\circ}$ then $x \in {^\circ\beta^\circ} \backslash {^\circ\beta}$ (since $x \in {^\circ(\beta \odot \gamma)^\circ} \backslash {^\circ(\beta \odot \gamma)}$), hence $\alpha^\circ \cap {^\circ\beta^\circ} \backslash {^\circ\beta} \neq \emptyset$ and by Definition 11.2 (of concatenation) we would have $\alpha \odot \beta = \Omega_{\mathbb{Y}}$. Thus $x \notin {^\circ\beta^\circ}$ which implies $x \in {^\circ\gamma^\circ} \backslash {^\circ\gamma}$ (because ${^\circ(\beta \odot \gamma)^\circ} = {^\circ\beta^\circ} \cup {^\circ\gamma^\circ}$). Since $x \in \alpha^\circ \wedge x \notin {^\circ\beta^\circ}$ implies $x \in (\alpha \odot \beta)^\circ$, then (d') holds.

Therefore, we proved (a')∨(b') ⇒ (c')∨(d'). Reverse implication is proved analogously. □

Theorem 11.2 (associativity of $\odot$ and notation: processes $\mathbb{O}$ and $\mathbb{I}$)

(a) $\alpha \odot (\beta \odot \gamma) = (\alpha \odot \beta) \odot \gamma$ *Concatenation of processes is associative*
(b) $\Omega_{\mathbb{Y}} \odot \alpha = \alpha \odot \Omega_{\mathbb{Y}} = \Omega_{\mathbb{Y}}$ *$\Omega_{\mathbb{Y}}$ is a zero process for concatenation, denoted by $\mathbb{O}$*
(c) $\theta \odot \alpha = \alpha \odot \theta = \alpha$ *θ is a unit (neutral) process for concatenation, denoted by $\mathbb{I}$*

Proof **Of (a)**. By Proposition 11.3 we limit ourselves to the case when both sides of the equality are distinct from $\Omega_{\mathbb{Y}}$. By Definition 11.2 (of concatenation): $\alpha \odot (\beta \odot \gamma) = \alpha + (\beta \odot \gamma)^\alpha = \alpha + (\beta + \gamma^\beta)^\alpha =$ (by (a) in Proposition 11.2) $\alpha + \beta^\alpha + (\gamma^\beta)^\alpha$. On the other hand, $(\alpha \odot \beta) \odot \gamma = \alpha \odot \beta + \gamma^{\alpha \odot \beta} = \alpha + \beta^\alpha + \gamma^{\alpha \odot \beta}$. Thus, what suffices to be shown is $(\gamma^\beta)^\alpha = \gamma^{\alpha \odot \beta}$. Applying Definition 11.1 (of shift) we conclude that $\gamma^{\alpha \odot \beta}$ is γ with each $< x, k > \in car(\gamma)$ replaced with $< x, k + \Psi_{\alpha \odot \beta}(x) >$. But: $\Psi_{\alpha \odot \beta}(x) = \Psi_{\alpha + \beta^\alpha}(x) =$ (by (b) in Proposition 11.2) $max(\Psi_\alpha(x), \Psi_{\beta^\alpha}(x)) =$ (by (c) in Proposition 11.2) $max(\Psi_\alpha(x), \Psi_\beta(x) + \Psi_\alpha(x)) = \Psi_\beta(x) + \Psi_\alpha(x)$. Hence, $\gamma^{\alpha \odot \beta}$ is γ with each $< x, k > \in car(\gamma)$ replaced with $< x, k + \Psi_\beta(x) + \Psi_\alpha(x) >$. Therefore $(\gamma^\beta)^\alpha = \gamma^{\alpha \odot \beta}$.

Of (b) and (c): directly from Definition 11.2. This ends the proof of the theorem. □

Corollary 11.1 *The algebra $\mathscr{M} =< \boldsymbol{PR}, \odot, \mathbb{I} >$ is a monoid of processes.* □

It is worth noting that the inductive construction of (finite) processes (Definition 10.6) yields a natural inductive definition of concatenation:

$$\alpha \odot \mathbb{I} = \alpha$$
$$\alpha \odot \mathbb{O} = \mathbb{O}$$
$$\alpha \odot \varepsilon = suc_\varepsilon(\alpha)$$
$$\alpha \odot suc_\varepsilon(\beta) = suc_\varepsilon(\alpha \odot \beta)$$

for $\alpha, \beta \in \boldsymbol{PR}$, $\varepsilon \in \boldsymbol{PR}_0$, ($\mathbb{I} \odot \alpha = \alpha$ and $\mathbb{O} \odot \alpha = \mathbb{O}$ follow from the construction of process). Associativity of $\odot$ follows directly from its inductive definition.

Due to associativity we write $\alpha \odot \beta \odot \gamma$ for $\alpha \odot (\beta \odot \gamma)$ or $(\alpha \odot \beta) \odot \gamma$ and this extends to any number of operands. But since processes may be infinite, it is appropriate to extend this also to infinity.

Definition 11.3 (*infinite concatenation*) For the infinite sequence of processes $\alpha_n \in \boldsymbol{PR},\ n \geq 1,$ the infinite concatenation is defined as:

$$\alpha_1 \odot \alpha_2 \odot .... \odot \alpha_n \odot .... = \sum_{n\geq 1} \alpha_1 \odot \alpha_2 \odot .... \odot \alpha_n$$

Recall that the infinite sum is the least upper bound—see Chap. 6. □

Note that $\alpha_1 \odot \alpha_2 \odot .... \odot \alpha_m \leq \alpha_1 \odot \alpha_2 \odot .... \odot \alpha_n$ for $m \leq n$ and that the least upper bound $\sum_{n\geq 1} \alpha_1 \odot \alpha_2 \odot .... \odot \alpha_n$ exists. Indeed, if $\alpha_1 \odot \alpha_2 \odot .... \odot \alpha_m = \mathbb{O}$ for a certain m then $\alpha_1 \odot \alpha_2 \odot .... \odot \alpha_n = \mathbb{O}$ for $n \geq m$ thus $\sum_{n\geq 1} \alpha_1 \odot \alpha_2 \odot .... \odot \alpha_n = \mathbb{O}$. Suppose $\alpha_1 \odot \alpha_2 \odot .... \odot \alpha_n \neq \mathbb{O}$ for each n. By construction of process, if a node $< x, k >$ occurs in $\alpha_1 \odot \alpha_2 \odot .... \odot \alpha_m$ then it occurs with the same superscript (cause) and subscript (effect) polynomials (monomials, in fact) in any process $\alpha_1 \odot \alpha_2 \odot .... \odot \alpha_n$ for $n \geq m$. Hence, the polynomials of $< x, k >$ do not grow in length in processes $\alpha_1 \odot \alpha_2 \odot .... \odot \alpha_n$ with growing n (their unbounded growth was the case shown in Fig. 6.1, Chap. 6). Therefore $\sum_{n\geq 1} \alpha_1 \odot \alpha_2 \odot .... \odot \alpha_n$ exists.

Applying (a) in Proposition 11.2, it may be checked that: $\alpha_1 \odot \alpha_2 \odot .... \odot \alpha_n \odot .... = \sum_{n\geq 1}(...(\alpha_n^{\alpha_{n-1}})^{\alpha_{n-2}}...)^{\alpha_1} = \alpha_1 + \alpha_2^{\alpha_1} + (\alpha_3^{\alpha_2})^{\alpha_1}...(...(\alpha_n^{\alpha_{n-1}})^{\alpha_{n-2}}...)^{\alpha_1}...$

Definition of concatenation and Theorem 11.2 determine the algebraic structure $\langle \boldsymbol{PR}, \odot, \mathbb{O}, \mathbb{I}\rangle$, that is a monoid with zero, which is the starting point to the theory of process languages. Some useful properties of this monoid are collected in:

Proposition 11.4 *(a) If $u \in \boldsymbol{FC}_{\mathbb{X}}^*$ and $v \in \boldsymbol{FC}_{\mathbb{X}}^* \cup \boldsymbol{FC}_{\mathbb{X}}^{\omega}$ then $pr(uv) = pr(u) \odot pr(v)$. This means that the mapping $pr\colon \boldsymbol{FC}_{\mathbb{X}}^* \cup \boldsymbol{FC}_{\mathbb{X}}^{\omega} \rightarrow \boldsymbol{PR}$ is a homomorphism of the partial (since the string concatenation is not defined for infinite left operand) monoid $\langle \boldsymbol{FC}_{\mathbb{X}}^* \cup \boldsymbol{FC}_{\mathbb{X}}^{\omega}, ., \varepsilon\rangle$ into the monoid $\langle \boldsymbol{PR}, \odot, \mathbb{I}\rangle$.*
(b) Any process is a finite or infinite concatenation of atomic processes, that is, $\mathbb{O}$ or $\mathbb{I}$ or $pr(Q)$, for an arbitrary firing component $Q \in \boldsymbol{FC}_{\mathbb{X}}$. If $\alpha = \varepsilon_1 \odot \varepsilon_2 \odot \varepsilon_3 \odot .... = \delta_1 \odot \delta_2 \odot \delta_3 \odot ....$ where ε_j and δ_j are atomic processes, then one of the sequences $(\varepsilon_1, \varepsilon_2, \varepsilon_3, ...)$, $(\delta_1, \delta_2, \delta_3, ...)$ is a permutation of the other. Two consecutive atoms $\varepsilon_j, \varepsilon_{j+1}$ may be swapped without affecting process α if and only if either $\varepsilon_j = \varepsilon_{j+1}$ or ${}^{\circ}\varepsilon_j^{\circ} \cap {}^{\circ}\varepsilon_{j+1}^{\circ} = \emptyset$.
(c) Let $u, u' \in \boldsymbol{FC}_{\mathbb{X}}^*$, $v, v' \in \boldsymbol{FC}_{\mathbb{X}}^* \cup \boldsymbol{FC}_{\mathbb{X}}^{\omega}$, $pr(u') = pr(u)$, $pr(v') = pr(v)$. Then $pr(u'v') = pr(uv)$. So, the equivalence $\overset{*}{\approx}$ in Theorem 10.4 is a congruence.
(d) For any processes $\alpha, \beta, \gamma, \delta$ such that $\alpha \odot \beta = \gamma \odot \delta$ there exist processes $\eta_1, \eta_2, \eta_3, \eta_4$, with $\alpha = \eta_1 \odot \eta_2$, $\beta = \eta_3 \odot \eta_4$ $\gamma = \eta_1 \odot \eta_3$, $\delta = \eta_2 \odot \eta_4$. This phenomenon is depicted in Fig. 11.2.
This property is the Levi Lemma [3] for processes.

Proof **Of (a)**. The equation holds if $u \notin \boldsymbol{FS}_{\mathbb{X}}$ or $v \notin \boldsymbol{FS}_{\mathbb{X}}$, since both sides of the equation are then $\mathbb{O}$. If $u \in \boldsymbol{FS}_{\mathbb{X}}$ and $v \in \boldsymbol{FS}_{\mathbb{X}}$ and $uv \notin \boldsymbol{FS}_{\mathbb{X}}$ then the equation holds too: by Theorem 10.2 its both sides are $\mathbb{O}$. Let:

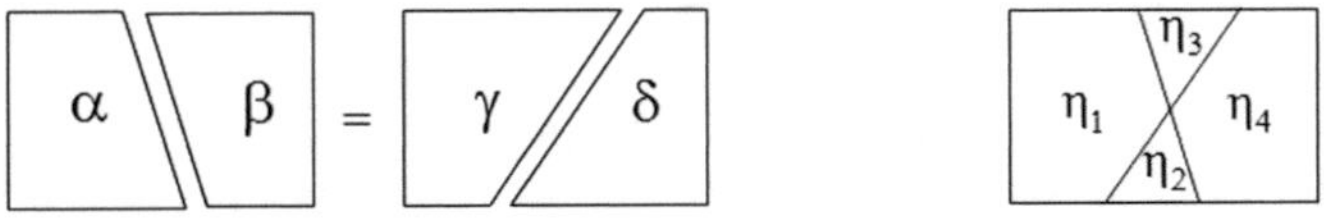

Fig. 11.2 Levi Lemma for processes

$$u = \underbrace{Q_1 Q_2 .... Q_j}_{u_j} .... Q_n \in \boldsymbol{FS}_{\mathbb{X}}, \qquad v = \underbrace{P_1 P_2 .... P_j}_{v_j} .... \in \boldsymbol{FS}_{\mathbb{X}}, \qquad uv \in \boldsymbol{FS}_{\mathbb{X}}.$$

Then, by Definition 10.3 (of process):

$$\underbrace{Q_1^{u_1} + Q_2^{u_2} + ... + Q_n^{u_n}}_{pr(u)} + \underbrace{P_1^{w_1} + P_2^{w_2} + ... + P_j^{w_j} ....}_{pr(v)^{pr(u)}} = pr(u) \odot pr(v)$$

where $w_j(x) = [u_n v_j](x)$ for $j \geq 1$.

Of (b). By (a): $pr(Q_1 Q_2 ....) = pr(Q_1) \odot pr(Q_2) \odot ....$ and by Theorem 11.2.

Of (c). Directly from (a).

Of (d). The proof for abstract monoids is in [3], for processes is in [2]. Indirect evidence follows also from Theorem 10.4 and the link between processes and Mazurkiewicz traces for which the Levi Lemma has been proved in [1]. □

References

1. Diekert V, Rozenberg G (eds) (1995) The book of traces. World Scientific, Singapore, New Jersey, London, Hong Kong
2. Kusmirek A (1996) Axiomatic characterisation and some properties of processes in cause-effect nets (Levi's lemma, pumping lemma and others). M.Sc. thesis (in Polish), Warsaw University
3. Levi FW (1944) On semigroups. Bull Calcuta Math Soc 36:141146

Chapter 12
Languages of Processes: Analysis and Synthesis Problems

A language of processes is any set of processes. Some of such languages represent behaviour of cause-effect structures and will be called definable (or generable) by c-e structures, "ces-definable", in short. Following the theory of formal languages of strings, we intend to find a definability of process languages by finite expressions in a suitably devised algebra, or, equivalently, by means of grammars or equations of a special form. Thus, we strive for linking the operational definability by c-e structures, with algebraic definability. So, there is a close analogy to the corresponding task in the theory of regular languages of strings, solved by respective Kleene theorem [4]. In accordance with Theorems 11.1 and 11.2, the set of all processes $\boldsymbol{PR}$ with concatenation is a monoid which, apart from the unity $\mathbb{I}$, contains a distinguished element $\mathbb{O}$. This algebra $\langle \boldsymbol{PR}, \odot, \mathbb{O}, \mathbb{I} \rangle$ yields the algebra of its subsets, that is, process languages, called sometimes ω-process languages [2], since they may contain also infinite processes. By $\mathbb{P}(\boldsymbol{PR})$ will be denoted the powerset of $\boldsymbol{PR}$, that is, the set of all subsets of $\boldsymbol{PR}$. For $A, B \subseteq \boldsymbol{PR}$, define:

$$A \odot B = \{\alpha \odot \beta : \alpha \in A \wedge \beta \in B\} \text{ and}$$

$$A^{\odot} = \bigcup_{n \geq 0} A^n \quad \text{where } A^0 = \{\mathbb{I}\}, \quad A^{n+1} = A^n \odot A (n \geq 0)$$

The algebra $\langle \mathbb{P}(\boldsymbol{PR}), \cup, \cap, \setminus, \odot, ^{\odot}, \{\mathbb{O}\}, \{\mathbb{I}\} \rangle$ is a counterpart of the algebra of string languages and a number of algebraic laws are readily transferable from the latter, but some laws are not. For instance, for any process languages $A, B \subseteq \boldsymbol{PR}$:

(1) $A \odot \{\mathbb{I}\} = \{\mathbb{I}\} \odot A = A$
(2) $A \odot \{\mathbb{O}\} = \{\mathbb{O}\} \odot A = \{\mathbb{O}\}$ for $A \neq \emptyset$
(3) $A \odot \emptyset = \emptyset \odot A = \emptyset$
(4) $\{\mathbb{I}\}^{\odot} = \emptyset^{\odot} = \{\mathbb{I}\}$
(5) $\{\mathbb{O}\}^{\odot} = \{\mathbb{O}, \mathbb{I}\}$
(6) $(A^{\odot} \cup B)^{\odot} = (A \cup B)^{\odot}$
(7) $A^{\odot} \odot A^{\odot} = A^{\odot}$

L. Czaja, *Cause-Effect Structures*, Lecture Notes in Networks and Systems 45, https://doi.org/10.1007/978-3-030-20461-7_12

(8) $\boldsymbol{PR} \odot \boldsymbol{PR} = \boldsymbol{PR}$
(9) $A \odot (B \odot C) = (A \odot B) \odot C$
(10) $A \odot (B \cup C) = A \odot B \cup A \odot C$
(11) $A \subseteq B \Rightarrow A \odot C \subseteq B \odot C \wedge C \odot A \subseteq C \odot B$
(12) $A \odot B = \emptyset \Rightarrow A = \emptyset \vee B = \emptyset$
(13) $A \odot B = \{\mathbb{I}\} \Rightarrow A = \{\mathbb{I}\} \wedge B = \{\mathbb{I}\}$
(14) The least (with respect to $\subseteq$) solution of the equation
$X = A \odot X \cup B$ is $X = A^{\odot} \odot B$
(15) The least solution of the equation $X = X \odot A \cup B$ is $X = B \odot A^{\odot}$
(16) The general solution of the equation $X = A \odot X \cup B$ is
$X = A^{\odot} \odot (B \cup \partial(A) \odot C)$
(17) The general solution of the equation $X = X \odot A \cup B$ is
$X = (B \cup \partial(A) \odot C) \odot A^{\odot}$
where C is an arbitrary process language and:

$$\partial(A) = \begin{cases} \{\mathbb{I}\} & \text{if } \mathbb{I} \in A \\ \emptyset & \text{if } \mathbb{I} \notin A \end{cases}$$

(let us assume that $\odot$ binds stronger than $\cup$)
(18) For any countable indexed family of process languages $\{A_i\}_{i \in I}$ and a process language B: $B \odot \bigcup_{i \in I} A_i = \bigcup_{i \in I} B \odot A_i$ (generalization of (10)).

The proofs are straightforward and are left as exercises. Note that equations in (14), (15) may be extended to systems of equations: where a vector X of process languages should be found. These equations are special cases of algebraic equations considered later in this chapter. Unlike in the algebra of string languages, for some non-empty and distinct from $\{\mathbb{O}\}$ and $\{\mathbb{I}\}$ process languages A, the language $A^{\odot}$ may happen to be finite, e.g. for $A = \{< a, 0 > \rightarrow < b, 1 >\}$, we have

$$A^{\odot} = \{\mathbb{O}, \mathbb{I}, < a, 0 > \rightarrow < b, 1 >\},$$

which shows that not every law for string languages, holds for process languages. $A^{\odot}$ is a set of finite concatenations of processes from A, that is:

$$A^{\odot} = \{\alpha \in \boldsymbol{PR} \,|\, there\, exist\, \alpha_1, \alpha_2, .., \alpha_n \in A\, with$$
$$\alpha = \alpha_1 \odot \alpha_2 \odot ... \odot \alpha_n\} \cup \{\mathbb{I}\} \tag{12.1}$$

For the purpose of adequate formulation of the analysis and synthesis problems, let us define the set $A^{(\omega)}$ of infinite concatenations of processes from the set A:

$$A^{(\omega)} = \{\alpha \in \boldsymbol{PR} |\; there\ exist\ \alpha_1, \alpha_2, ... \in A\ (infinitely\ many)\ with$$
$$\alpha = \alpha_1 \odot \alpha_2 \odot ...\} \tag{12.2}$$

For infinite concatenation of processes—see Definition 11.3. Note that if $\mathbb{I} \in A$ then $A^{\odot} \subseteq A^{(\omega)}$. Indeed, each concatenation $\alpha_1 \odot \alpha_2 \odot ... \odot \alpha_n$ may be seen as:

$$\alpha_1 \odot \alpha_2 \odot ... \odot \alpha_n \odot \underbrace{\mathbb{I} \odot \mathbb{I} \odot \mathbb{I} \odot \mathbb{I} \odot \mathbb{I} \odot ....}_{infinite\ concatenation}$$

Let us also introduce denotation for the set of finite or infinite concatenations of processes from A:

$$A^{(\infty)} = A^{\odot} \cup A^{(\omega)} \tag{12.3}$$

Notice that iteration operators in parentheses $^{(\omega)}$ and $^{(\infty)}$ are based on concatenation $\odot$ of processes, whereas not-in-parentheses $^{\omega}$ and $^{\infty}$ are based on concatenation of sequences (words).

Considering $^{\odot}$ and $^{(\omega)}$ and $^{(\infty)}$ as unary operators (of iteration) on process languages, we come to:

Definition 12.1 (*Algebra of Process Languages*) *Algebra of Process Languages (APL)* is the power monoid of processes, supplemented with operators $\cup, \cap, \setminus, ^{\odot}, ^{(\omega)}, ^{(\infty)}$ and with distinguished singleton set $\{\mathbb{O}\}$, that is:

$$APL = \langle \mathbb{P}(\boldsymbol{PR}), \cup, \cap, \setminus, \odot, ^{\odot}, ^{(\omega)}, ^{(\infty)}, \{\mathbb{O}\}, \{\mathbb{I}\} \rangle$$ □

Now, the analysis and synthesis problems may be formulated and solved. These problems, probably the most significant in the theory of formal (string) languages and automata, have almost immediate solutions for process languages. Recall that processes and their sets, i.e. process languages, have been defined regardless of any generating devices. But c-e structures, in a natural way, are such devices. The synthesis problem consists in finding such c-e structure, given a process language, while analysis is a converse task. Let us make this precise.

Definition 12.2 (*process languages generated by c-e structures*) Let $U \in \boldsymbol{CE}[\mathbb{X}]$ be a c-e structure and let $s_0 \subseteq \mathbb{X}$ be a state. Denote: $\boldsymbol{PR}[U] = \{pr(u) \mid u \in \boldsymbol{FC}_{\mathbb{X}}[U]^{\infty}\}$, where $\boldsymbol{FC}_{\mathbb{X}}[U]^{\infty} = \boldsymbol{FC}_{\mathbb{X}}[U]^{*} \cup \boldsymbol{FC}_{\mathbb{X}}[U]^{\omega}$ and where $\boldsymbol{FC}_{\mathbb{X}}[U]^{*}$ and $\boldsymbol{FC}_{\mathbb{X}}[U]^{\omega}$ are, respectively, the set of all finite and infinite sequences of firing components of U. We say that $\boldsymbol{PR}[U]$ is the *process language generated by a c-e structure* U. Denote: $\boldsymbol{PR}[U, s_0] = \{pr(u) \mid u \in \boldsymbol{FC}_{\mathbb{X}}[U]^{\infty} \wedge$ *there is a u-computation* $s_0, s_1, s_2, \ldots$ (Definition 10.1)$\}$

Say that $\boldsymbol{PR}[U, s_0]$ is the *process language generated by U starting from initial state* s_0. □

12.1 Analysis

The analysis problems are stated as:
(A1) Given c-e structure $U \in \boldsymbol{CE}[\mathbb{X}]$ find a process language $\boldsymbol{PR}[U]$
(A2) Given c-e structure $U \in \boldsymbol{CE}[\mathbb{X}]$ with initial state $s_0 \subseteq \mathbb{X}$, find a process language $\boldsymbol{PR}[U, s_0]$
One would like, however, to have these formulations more specific, since it is not known how the c-e structure U is given and how the process language is to be

presented. We shall assume that the c-e structure U is given by an arrow expression in the quasi-semiring of c-e structures or by a set of nodes superscripted and subscripted by polynomials (Chap. 2) and the process language is to be presented by a finite expression in the algebra APL with atomic processes (see (b) in Proposition 11.4) as operands. A solution to the problems (A1) and (A2) is provided by:

Theorem 12.1 (analysis—first solution) *For each c-e structure* $U \in \boldsymbol{CE}[\mathbb{X}]$ *and state* $s_0 \subseteq \mathbb{X}$*:*
(AA1) $\boldsymbol{PR}[U] = \{pr(Q) \mid Q \in \boldsymbol{FC}_{\mathbb{X}}[U]\}^{(\infty)}$
(AA2) $\boldsymbol{PR}[U, s_0] = \boldsymbol{PR}[U] \cap PREF[\{\alpha \in \boldsymbol{PR} \mid {}^\circ\alpha = s_0\}]$
where, for a given process language A,

$$PREF[A] = \bigcup_{\alpha \in A} \{\beta \in \boldsymbol{PR} \mid \beta \, is \; a \; prefix \; of \alpha\}$$

and, obviously, β *is a prefix of* α *iff* $\alpha = \beta \odot \gamma$ *for a certain* γ*; as in previous chapters,* ${}^\circ\alpha$ *is a set of minimal nodes of* α *projected onto* $\mathbb{X}$.

Proof **Of (AA1).** $\boldsymbol{PR}[U] = \{pr(u) \mid u \in \boldsymbol{FC}_{\mathbb{X}}[U]^{\infty}\} = \{\alpha \in \boldsymbol{PR} \mid \exists u \in \boldsymbol{FC}_{\mathbb{X}}[U]^{\infty} : \alpha = pr(u)\} =$
$\{\alpha \in \boldsymbol{PR} \mid \exists Q_1, Q_2, ... \in \boldsymbol{FC}_{\mathbb{X}}[U] : (\alpha = pr(Q_1 Q_2 ...)\} =$ (by (a), (b) in Proposition 11.4) $=$
$\{\alpha \in \boldsymbol{PR} \mid \exists Q_1, Q_2, ... \in \boldsymbol{FC}_{\mathbb{X}}[U] : \alpha = pr(Q_1) \odot pr(Q_2) \odot ...\} =$
$\{\alpha \in \boldsymbol{PR} \mid \exists \alpha_1, \alpha_2, ... \in \{pr(Q) \mid Q \in \boldsymbol{FC}_{\mathbb{X}}[U]\} : \alpha = \alpha_1 \odot \alpha_2 \odot ...\} =$ (by (12.1), (12.2), (12.3))
$\{pr(Q) \mid Q \in \boldsymbol{FC}_{\mathbb{X}}[U]\}^{(\infty)}$. Notice that α_j are atomic processes.
Of (AA2) $\boldsymbol{PR}[U, s_0] = \{pr(u) \mid u \in \boldsymbol{FC}_{\mathbb{X}}[U]^{\infty} \wedge$ *there is a u-computation* $s_0, s_1, s_2, ...\} =$
$\{pr(u) \mid u \in \boldsymbol{FC}_{\mathbb{X}}[U]^{\infty}\} \cap \{pr(u) \mid$ *there is a u-computation* $s_0, s_1, s_2, ...\} =$
$\boldsymbol{PR}[U] \cap PREF[\{\alpha \in \boldsymbol{PR} \mid {}^\circ\alpha = s_0\}]$. □

Proposition 12.1 *For any* $U \in \boldsymbol{CE}[\mathbb{X}]$*:* $\boldsymbol{PR}\,[U] = \bigcup_{s \subseteq car(U)} \boldsymbol{PR}[U, s]$.

Proof By (AA2) in Theorem 12.1: $\bigcup_{s \subseteq car(U)} \boldsymbol{PR}[U, s] =$
$\bigcup_{s \subseteq car(U)} \boldsymbol{PR}[U] \cap PREF[\{\alpha \in \boldsymbol{PR} \mid {}^\circ\alpha = s\}] =$
$\boldsymbol{PR}[U] \cap \bigcup_{s \subseteq car(U)} PREF[\{\alpha \in \boldsymbol{PR} \mid {}^\circ\alpha = s\}] = \boldsymbol{PR}[U]$
because $\boldsymbol{PR}[U] \subseteq \bigcup_{s \subseteq car(U)} PREF[\{\alpha \in \boldsymbol{PR} \mid {}^\circ\alpha = s\}]$. □

Let us illustrate the analysis by:

Example 12.1 Consider Fig. 8.2 in Chap. 8. The c-e structure U has the firing components:
$Q_1 = (a \to u)$, $Q_2 = (b \to x)$, $Q_3 = (u \to c) \bullet (x \to c)$, $Q_4 = (c \to a) \bullet (c \to b)$

and the c-e structure V has the firing components:
$Q_5 = (a \rightarrow v)$, $Q_6 = (b \rightarrow y)$, $Q_7 = (v \rightarrow d) \bullet (y \rightarrow d)$, $Q_8 = (d \rightarrow a) \bullet (d \rightarrow b)$
$\boldsymbol{FC}_{\mathbb{X}}[U] = \{Q_1, Q_2, Q_3, Q_4\}$, $\boldsymbol{FC}_{\mathbb{X}}[V] = \{Q_5, Q_6, Q_7, Q_8\}$, so,
$\boldsymbol{FC}_{\mathbb{X}}[U+V] = \boldsymbol{FC}_{\mathbb{X}}[U] \cup \boldsymbol{FC}_{\mathbb{X}}[V]$ (no new firing components are created in making $U+V$), thus, by (b) in Proposition 8.1: $[[U]] \cup [[V]] = [[U+V]]$.
Atomic processes:
$pr(Q_1) = < a, 0 > \rightarrow < u, 1 >$, $pr(Q_2) = < b, 0 > \rightarrow < x, 1 >$,
$pr(Q_3) = (< u, 0 > \rightarrow < c, 1 >) \bullet (< x, 0 > \rightarrow < c, 1 >)$
$pr(Q_4) = (< c, 0 > \rightarrow < a, 1 >) \bullet (< c, 0 > \rightarrow < b, 1 >)$
$pr(Q_5) = < a, 0 > \rightarrow < v, 1 >$, $pr(Q_6) = < b, 0 > \rightarrow < y, 1 >$,
$pr(Q_7) = (< v, 0 > \rightarrow < d, 1 >) \bullet (< y, 0 > \rightarrow < d, 1 >)$
$pr(Q_8) = (< d, 0 > \rightarrow < a, 1 >) \bullet (< d, 0 > \rightarrow < b, 1 >)$

Notice that operation "$\bullet$" in above expressions represent multiplication of c-e structures, not processes.
By Theorem 12.1: $\boldsymbol{PR}[U+V] = \{pr(Q_1), pr(Q_2), pr(Q_3), pr(Q_4), pr(Q_5), pr(Q_6), pr(Q_7), pr(Q_8)\}^{(\infty)}$
Obviously, this is an expression in the *APL* defining a language of processes evoked by c-e structure $U+V$. □
Calling "*APL*-algebraic" a process language representable by a finite expression in the algebra *APL* with atomic processes as operands, we have:

Corollary 12.1 *Any process language generated by a finite c-e structure is APL-algebraic.* □
We need some auxiliary notions for the forthcoming considerations:

Definition 12.3 (*folding $FOLD$*) Let $\alpha \in \boldsymbol{PR}$ and let $\alpha = pr(u)$ with $u = Q_1 Q_2 Q_3 \ldots\ldots$ ($Q_j \in \boldsymbol{FC}_{\mathbb{X}}$). *Folding* of process α is defined as: $FOLD[\alpha] = \sum_{j \geq 1} Q_j$ for $u \neq \varepsilon$ and $FOLD[\alpha] = \theta$ for $u = \varepsilon$. Additionally, let $FOLD[\mathbb{O}] = FOLD[\mathbb{I}] = \theta$. That is, folding of α is the sum of its firing components projected onto $\mathbb{X}$ (for instance, projection of $< a, 0 > \rightarrow < b, 1 >$ onto $\mathbb{X}$ is $a \rightarrow b$). □

It may be verified that any finite process is a correct c-e structure over $\mathbb{Y} = \mathbb{X} \times \mathbb{N}$ and if $\alpha \in \boldsymbol{PR}[U]$ for a certain c-e structure U, then $FOLD[\alpha] \leq U$, that is, $FOLD[\alpha]$ is a substructure of U.

Definition 12.4 (*function Δ_α*) Let $\alpha \in \boldsymbol{PR}$ and let $\alpha = pr(u)$ with $u = Q_1 Q_2 Q_3 \ldots\ldots$ ($Q_j \in \boldsymbol{FC}_{\mathbb{X}}$). To the process α, an integer-valued function Δ_α with arguments from $\mathbb{X}$ is associated and defined as: $\Delta_\alpha(x) = \sum_{j \geq 1} (\chi_{\bullet Q_j}(x) - \chi_{Q_j^\bullet}(x))$ where χ_A is the characteristic function of the set A. Additionally let $\Delta_\alpha(x) = 0$ for $\alpha = \mathbb{O}$ or $\alpha = \mathbb{I}$. □

Obviously, $\Delta_\alpha(x)$ is the difference between the number of arrows going out of and coming into all nodes $< x, k >$ ($0 \leq k \leq \Psi_\alpha(x)$ or $1 \leq k \leq \Psi_\alpha(x)$) in the process α. Note that Δ_α may assume three values only: -1, 0 or $+1$ (because in elementary c-e structures any place x may hold one token at the most).

12.2 Other Solutions to the Analysis Problem

Theorem 12.2 (analysis - second solution) *For each c-e structure* $U \in \boldsymbol{CE}[\mathbb{X}]$ *and state* $s_0 \subseteq \mathbb{X}$*:*
(AAA1) $\boldsymbol{PR}[U] = \{\alpha \in \boldsymbol{PR} | \ FOLD[\alpha] \leq U\}$
(AAA2) The sets $A = \{\alpha \in \boldsymbol{PR} \mid \ ^\circ\alpha = s_0\}$ *in (AA2) and*
$B = \{\alpha \in \boldsymbol{PR} \mid \ \forall x : (x \in s_0 \Leftrightarrow \exists \beta \in PREF[\{\alpha\}] : \Delta_\beta(x) = 1)\}$ *are identical.*

Proof **Of (AAA1)**: evident, since $FOLD[\alpha] \leq U$ if and only if $\alpha \in \boldsymbol{PR}[U]$.
Of (AAA2). Let a process α with $^\circ\alpha = s_0$ be given. It is evident that $x \in {^\circ\alpha}$ iff $\Delta_\beta(x) = 1$ for a certain prefix β of α. Thus $A \subseteq B$.

Assume that α satisfies $\forall x : (x \in s_0 \Leftrightarrow \exists \beta \in PREF[\{\alpha\}] : \Delta_\beta(x) = 1)$. If $x \in s_0$ then (by ($\Rightarrow$) in the latter assumption) $\Delta_\beta(x) = 1$ for a certain prefix β of α, thus $x \in {^\circ\beta}$, which implies $x \in {^\circ\alpha}$. Thus $s_0 \subseteq {^\circ\alpha}$. If $x \in {^\circ\alpha}$ then $\Delta_\beta(x) = 1$ for a certain prefix β of α, thus (by ($\Leftarrow$) in the assumption) $x \in s_0$. Hence $^\circ\alpha \subseteq s_0$. Therefore $^\circ\alpha = s_0$ thus $B \subseteq A$. Concluding: $A = B$. □

To provide still another solution to the analysis problem, let us define the so called *incidence matrix* of a given c-e structure U, which we denote by $\underline{U}$. (in the Petri nets context—cf. [5]). Its rows correspond to the nodes of U, its columns—to the firing components of U (that is, we assume some orders in the sets $car(U)$ and $\boldsymbol{FC}_{\mathbb{X}}[U]$). For $x \in car(U)$ and $Q \in \boldsymbol{FC}_{\mathbb{X}}[U]$ the (x, Q)-entry of $\underline{U}$ is defined as:

$$\underline{U}[x, Q] = \begin{cases} +1 \text{ if } x \in Q^\bullet \\ -1 \text{ if } x \in {^\bullet Q} \\ 0 \quad \text{else} \end{cases}$$

For a process $\alpha = pr(Q_1Q_2Q_3.....) \in \boldsymbol{PR}[U]$, distinct from $\mathbb{O}$ and $\mathbb{I}$, let us denote by $\underline{\alpha}$ a vector (called a Parikh vector of α) whose constituents $\underline{\alpha}\ [Q]$ correspond to firing components of U ordered as columns in the matrix $\underline{U}$. The Q-entry of $\underline{\alpha}$ is defined as:

$\underline{\alpha}[Q]$ = number of indices j in the sequence $Q_1Q_2Q_3.....$ such that $Q_j = Q$.

It is possible that $\underline{\alpha}\ [Q] = \infty$. Let us assume the following arithmetics of ∞: $\infty + \infty = \infty$, $\infty - \infty = 0$, $\infty \cdot 0 = 0$, $\infty \cdot k = \infty$, for any integer $k > 0$ and $\infty \cdot k = -\infty$, for $k < 0$. That is, number $\underline{\alpha}\ [Q]$ says how many times Q fires during evolution of the process α. For example, if U is the "triangle" c-e structure in Fig. 10.2 and α is the process in Fig. 10.4 (Chap. 10) then:

$\underline{U}$ =

	$a \to b$	$b \to c$	$c \to a$
a	-1	0	$+1$
b	$+1$	-1	0
c	0	$+1$	-1

$\underline{\alpha}$ =

$a \to b$	$b \to c$	$c \to a$
3	2	2

Proposition 12.2 *Let a c-e structure* U *and a process* $\alpha \in \boldsymbol{PR}[U]$ *be given. Then* $\underline{U} \cdot \underline{\alpha} = -\Delta_\alpha$.

Here, function Δ_α *is treated as a vector (or multiset over* $\mathbb{X}$*) whose constituents* $\Delta_\alpha(x)$ *correspond to nodes* $x \in car(U)$ *ordered as rows in the matrix* $\underline{U}$.

Proof Let $\alpha = pr(Q_1Q_2Q_3.....)$ with $Q_j \in \boldsymbol{FC}_{\mathbb{X}}[U]$. By definition of $\underline{U}$: $\underline{U}\,[x, Q] = \chi_{Q^\bullet}(x) - \chi_{^\bullet Q}(x)$, thus the inner product of x's row of $\underline{U}$ by $\underline{\alpha}$, i.e. $\sum_{Q \in \boldsymbol{FC}_{\mathbb{X}}[U]} \underline{U}\,[x, Q] \cdot \underline{\alpha}[Q]$, equals $\sum_{Q \in \boldsymbol{FC}_{\mathbb{X}}[U]} (\chi_{Q^\bullet}(x) \cdot \underline{\alpha}[Q] - \chi_{^\bullet Q}(x) \cdot \underline{\alpha}[Q])$. By definition of $\underline{\alpha}$:

$I_\alpha(x) = \sum_{Q \in \boldsymbol{FC}_{\mathbb{X}}[U]} \chi_{Q^\bullet}(x) \cdot \underline{\alpha}[Q]$ number of arrows coming *I*nto all nodes $< x, k >$ in α

$O_\alpha(x) = \sum_{Q \in \boldsymbol{FC}_{\mathbb{X}}[U]} \chi_{^\bullet Q}(x) \cdot \underline{\alpha}[Q]$ number of arrows going *O*ut of all nodes $< x, k >$ in α

Thus, by Definition 12.4: $I_\alpha(x) - O_\alpha(x) = -\Delta_\alpha(x)$. If $\underline{\alpha}[Q] < \infty$ then $I_\alpha(x) - O_\alpha(x) = \sum_{Q \in \boldsymbol{FC}_{\mathbb{X}}[U]} (\chi_{Q^\bullet}(x) \cdot \underline{\alpha}[Q] - \chi_{^\bullet Q}(x) \cdot \underline{\alpha}[Q])$. If $\underline{\alpha}[Q] = \infty$ then $I_\alpha(x) = \infty$ for $x \in Q^\bullet$. But then also $O_\alpha(x) = \infty$ because α is a correct process, thus for each finite prefix β of α: $|I_\beta(x) - O_\beta(x)| \leq 1$. Similarly, $O_\alpha(x) = \infty$ implies $I_\alpha(x) = \infty$. Thus, by assumption $\infty - \infty = 0$, also in this case

$$I_\alpha(x) - O_\alpha(x) = \sum_{Q \in \boldsymbol{FC}_{\mathbb{X}}[U]} (\chi_{Q^\bullet}(x) \cdot \underline{\alpha}[Q] - \chi_{^\bullet Q}(x) \cdot \underline{\alpha}[Q]).$$

Therefore, $\sum_{Q \in \boldsymbol{FC}_{\mathbb{X}}[U]} \underline{U}\,[x, Q] \cdot \underline{\alpha}[Q] = -\Delta_\alpha(x)$ for each x, that is $\underline{U} \cdot \underline{\alpha} = -\Delta_\alpha$.

□

Theorem 12.2 and Proposition 12.2 readily imply:

Theorem 12.3 (analysis—third solution) *(AAAA1) as (AA1) in Theorem 12.1 or (AAA1) in Theorem 12.2.*
(AAAA2) $\boldsymbol{PR}[U, s_0] =$
$PREF[\{\alpha \in \boldsymbol{PR}[U] |\ \forall x : (x \in s_0 \Leftrightarrow \exists \beta \in PREF[\{\alpha\}] : (\underline{U} \cdot \underline{\beta})[x] = -1)\}]$. □

12.3 Synthesis

For making the notation shorter, denote for $A, B \subseteq \boldsymbol{PR}$: $A \simeq B$ iff $A \backslash \{\mathbb{O}, \mathbb{I}\} = B \backslash \{\mathbb{O}, \mathbb{I}\}$, that is iff process languages A and B differ by $\mathbb{O}$ or $\mathbb{I}$ at the most. The synthesis problems are:

(S1) Given a process language $L \subseteq \boldsymbol{PR}$, decide whether there exists a c-e structure $U \in \boldsymbol{CE}[\mathbb{X}]$ with $\boldsymbol{PR}[U] \simeq L$ (say then that L is *ces-definable*) and—in the positive case - find such U.

(S2) Given a process language $L \subseteq \boldsymbol{PR}$, decide whether there exists a c-e structure $U \in \boldsymbol{CE}[\mathbb{X}]$ and a state $s_0 \subseteq \mathbb{X}$ with $\boldsymbol{PR}[U, s_0] \simeq L$ (say then that L is *ces-init-definable*) and - in the positive case - find such U and s_0.

Proposition 12.3 *Let a process language* $L \subseteq \boldsymbol{PR}$ *be given. Then, for any c-e structure* $V \in \boldsymbol{CE}[\mathbb{X}]$:
If $\boldsymbol{PR}[V] \simeq L$ *then* $\boldsymbol{FC}_{\mathbb{X}}[V] = \boldsymbol{FC}_{\mathbb{X}}[\sum_{\alpha \in L} FOLD[\alpha]]$.

Proof Since $\alpha \in \boldsymbol{PR}[V]$ implies $FOLD[\alpha] \leq V$, then, by (c) in Proposition 6.2: $\sum_{\alpha \in \boldsymbol{PR}[V]} FOLD[\alpha] \leq V$ which, by (c) in Proposition 2.3, yields $\boldsymbol{FC}_{\mathbb{X}}[\sum_{\alpha \in \boldsymbol{PR}[V]} FOLD[\alpha]] \subseteq \boldsymbol{FC}_{\mathbb{X}}[V]$. Suppose $Q \in \boldsymbol{FC}_{\mathbb{X}}[V]$.

Obviously $FOLD[pr(Q)] = Q$, hence $Q \leq \sum_{\alpha \in \boldsymbol{PR}[V]} FOLD[\alpha]$. Since Q is a firing component, we get $Q \in \boldsymbol{FC}_{\mathbb{X}}[\sum_{\alpha \in \boldsymbol{PR}[V]} FOLD[\alpha]]$. Thus $\boldsymbol{FC}_{\mathbb{X}}[V] \subseteq FC_{\mathbb{X}}[\sum_{\alpha \in \boldsymbol{PR}[V]} FOLD[\alpha]]$.

Therefore $\boldsymbol{FC}_{\mathbb{X}}[V] = \boldsymbol{FC}_{\mathbb{X}}[\sum_{\alpha \in \boldsymbol{PR}[V]} FOLD[\alpha]]$. Applying assumption $\boldsymbol{PR}[V] \simeq L$ we get
$\boldsymbol{FC}_{\mathbb{X}}[V] = \boldsymbol{FC}_{\mathbb{X}}[\sum_{\alpha \in L} FOLD[\alpha]]$, because (on account of $FOLD[\mathbb{O}] = FOLD[\mathbb{I}] = \theta$)
if $L \simeq L'$ then $\sum_{\alpha \in L} FOLD[\alpha] = \sum_{\alpha \in L'} FOLD[\alpha]$. □

It should be noticed that not for every process language L the least upper bound $\sum_{\alpha \in L} FOLD[\alpha]$ exists. For instance, it does not exist for $L = \{< a, 0 > \to < \mathbf{1}, 1 >, < a, 0 > \to < \mathbf{2}, 1 >, < a, 0 > \to < \mathbf{3}, 1 >, ...\}$ since $FOLD[< a, 0 > \to < \mathbf{k}, 1 >] = a \to \mathbf{k}$ $(\mathbf{k} = \mathbf{1}, \mathbf{2}, \mathbf{3}, ...)$ - see beginning of Chap. 6.

Some solutions to the problems (S1) and (S2) are provided by:

Theorem 12.4 (synthesis) *Let a process language* $L \subseteq \boldsymbol{PR}$ *be given.*
(SS1) If L is ces-definable then $L \simeq \boldsymbol{PR}[U]$ *where* $U = \sum_{\alpha \in L} FOLD[\alpha]$
(SS2) If L is ces-init-definable then $L \simeq \boldsymbol{PR}[U, s_0]$ *where* $U = \sum_{\alpha \in L} FOLD[\alpha]$ *and* $s_0 = \bigcup_{\alpha \in L} {}^{\circ}\alpha$
Moreover, U is the least (w.r.t. $\leq$*) c-e structure generating L; in the case (SS2),* s_0 *is the least (w.r.t.* $\subseteq$*) state such that U starting from* s_0 *generates L.*

Proof **Of (SS1)**. Since L is ces-definable, then $L \simeq \boldsymbol{PR}[V]$ for a certain $V \in CE[\mathbb{X}]$. Thus, by Proposition 12.3: $\boldsymbol{FC}_{\mathbb{X}}[V] = \boldsymbol{FC}_{\mathbb{X}}[\sum_{\alpha \in L} FOLD[\alpha]]$. By (AA1) in Theorem 12.1: $\boldsymbol{PR}[V] = \{pr(Q) \mid Q \in \boldsymbol{FC}_{\mathbb{X}}[V]\}^{(\infty)}$, thus $\boldsymbol{PR}[V] = \{pr(Q) \mid Q \in \boldsymbol{FC}_{\mathbb{X}}[\sum_{\alpha \in L} FOLD[\alpha]]\}^{(\infty)} =$ (again by (AA1) in Theorem 12.1) $\boldsymbol{PR}[\sum_{\alpha \in L} FOLD[\alpha]]$. Therefore $L \simeq \boldsymbol{PR}[U]$ with $U = \sum_{\alpha \in L} FOLD[\alpha]$.

Of (SS2). Since L is ces-init-definable, then $L \backsimeq \boldsymbol{PR}[V, t_0]$ for a certain $V \in \boldsymbol{CE}[\mathbb{X}]$ and a certain state $t_0 \subseteq \mathbb{X}$. Thus, by (AA2) in Theorem 12.1:

$$L \backsimeq \boldsymbol{PR}[V] \cap PREF[\{\beta \in \boldsymbol{PR} | {}^\circ\beta = t_0\}].$$

We show that

$$L \backsimeq \boldsymbol{PR}[\sum_{\alpha \in L} FOLD[\alpha]] \cap PREF[\{\beta \in \boldsymbol{PR} | {}^\circ\beta = \bigcup_{\alpha \in L} {}^\circ\alpha\}].$$

Let $\gamma = pr(Q_1 Q_2 Q_3 ....) \in L$ and $\gamma \neq \mathbb{O}$, $\gamma \neq \mathbb{I}$. Thus, $\gamma \in PR[V]$ and $\gamma \in PREF[\{\beta \in \boldsymbol{PR} \mid {}^\circ\beta = t_0\}]$. Suppose $\gamma \notin \boldsymbol{PR}[\sum_{\alpha \in L} FOLD[\alpha]]$. Then there exists $Q \in \boldsymbol{FC}_{\mathbb{X}}[V]$ with $Q \notin \boldsymbol{FC}_{\mathbb{X}}[\sum_{\alpha \in L} FOLD[\alpha]]$. But this is in contradiction with Proposition 12.3, hence $\gamma \in \boldsymbol{PR}[\sum_{\alpha \in L} FOLD[\alpha]]$. From $\gamma \in L$ we infer ${}^\circ\gamma \subseteq \bigcup_{\alpha \in L} {}^\circ\alpha$, hence $\gamma \in PREF[\{\beta \in \boldsymbol{PR} \mid {}^\circ\beta = \bigcup_{\alpha \in L} {}^\circ\alpha\}]$. Therefore $L \subseteq \boldsymbol{PR}[\sum_{\alpha \in L} FOLD[\alpha]] \cap PREF[\{\beta \in \boldsymbol{PR} \mid {}^\circ\beta = \bigcup_{\alpha \in L} {}^\circ\alpha\}]$. The reverse inclusion is evident: it follows from $\boldsymbol{PR}[\sum_{\alpha \in L} FOLD[\alpha]] \subseteq \boldsymbol{PR}[V]$ and $PREF[\{\beta \in \boldsymbol{PR} \mid {}^\circ\beta = \bigcup_{\alpha \in L} {}^\circ\alpha\}] \subseteq PREF[\{\beta \in \boldsymbol{PR} \mid {}^\circ\beta = t_0\}]$. The two latter inclusions follow from $L \backsimeq \boldsymbol{PR}[V] \cap PREF[\{\beta \in \boldsymbol{PR} \mid {}^\circ\beta = t_0\}]$. Finally we get
$L \backsimeq \boldsymbol{PR}[\sum_{\alpha \in L} FOLD[\alpha]] \cap PREF[\{\beta \in \boldsymbol{PR} \mid {}^\circ\beta = \bigcup_{\alpha \in L} {}^\circ\alpha\}]$.
We proved (SS1) and (SS2). To prove that $U = \sum_{\alpha \in L} FOLD[\alpha]$ is the least (w.r.t $\leq$) c-e structure generating L, it suffices to recall that $FOLD[\alpha] \leq V$ for any $\alpha \in L$ and V satisfying $L \backsimeq \boldsymbol{PR}[V]$. Hence $U = \sum_{\alpha \in L} FOLD[\alpha] \leq V$. Demonstration that $\bigcup_{\alpha \in L} {}^\circ\alpha$ is included in any state s satisfying $L \backsimeq \boldsymbol{PR}[U] \cap PREF[\{\alpha \in \boldsymbol{PR} \mid {}^\circ\alpha = s\}]$ is left as an exercise. □

Example 12.2 Theorem 12.4 implies that the process language
$L = \{< a, 0 > \rightarrow < b, 1 >, < b, 0 > \rightarrow < a, 1 >\}$ is not ces-definable, since $U = \sum_{\alpha \in L} FOLD[\alpha]$ generates much more processes than L contains. Indeed:
$U = \{a \rightarrow b \rightarrow a\}$, thus $\boldsymbol{PR}\,[U] = \{\mathbb{O}, \mathbb{I}\} \cup$
$\{< a, 0 > \rightarrow < b, 1 > \rightarrow < a, 1 > \rightarrow < b, 2 > \rightarrow < a, 2 > \rightarrow < b, 3 > \rightarrow$ ...., etc.$\} \cup$
$\{< b, 0 > \rightarrow < a, 1 > \rightarrow < b, 1 > \rightarrow < a, 2 > \rightarrow < b, 2 > \rightarrow < a, 3 > \rightarrow$ ...., etc.$\} \cup$
{all finite prefixes of the two above infinite processes}.
Hence, $\boldsymbol{PR}[U] \backslash \{\mathbb{O}, \mathbb{I}\} \neq L \backslash \{\mathbb{O}, \mathbb{I}\}$. □

Example 12.3 Let L be a process language consisting of the disconnected infinite process (see Example 10.1 in Chap. 10):

$$\beta = \begin{array}{l} < a, 0 > \rightarrow < b, 1 > \rightarrow < c, 1 > \rightarrow < a, 2 > \rightarrow < b, 3 > \rightarrow < c, 3 > \rightarrow \;.... \\ < c, 0 > \rightarrow < a, 1 > \rightarrow < b, 2 > \rightarrow < c, 2 > \rightarrow < a, 3 > \rightarrow < b, 4 > \rightarrow \;.... \end{array}$$

along with all its finite prefixes. L is not ces-definable, but it is ces-init-definable. Indeed, $L \backsimeq \boldsymbol{PR}[U, s_0] = \boldsymbol{PR}[U] \cap PREF[\{\alpha \in \boldsymbol{PR} \mid {}^\circ\alpha = s_0\}]$, where U is the c-e structure in Example 10.1 and $s_0 = \{a, c\}$. □

Theorem 12.4 allows to compute a minimal c-e structure generating a given process language L under condition that the language has a generative c-e structures at all. But how such a condition looks like? A certain result, concerning c-e structures without initial state, provides the following:

Theorem 12.5 *Suppose that a process language L satisfies:*

$$\bigcup_{\alpha \in L} \boldsymbol{FC}_{\mathbb{X}}[FOLD[\alpha]] = \boldsymbol{FC}_{\mathbb{X}}[\sum_{\alpha \in L} FOLD[\alpha]] \tag{12.4}$$

that is, the summation $\sum$ does not create new firing components ((b) in Proposition 8.3). Then, L is ces-definable if and only if $L \backsimeq (L \cap \boldsymbol{PR}_0)^{(\omega)}$, where $\boldsymbol{PR}_0$ is the set of all atomic processes (Definition 10.5).

Proof Note that for any process language $A \subseteq \boldsymbol{PR}$:

$$A^{(\omega)} \cap \boldsymbol{PR}_0 = A \cap \boldsymbol{PR}_0. \tag{12.5}$$

Indeed, by (12.1) and (12.2): $A^{(\infty)} =$

$\{\alpha \in \boldsymbol{PR} \mid$ *there exist* $\alpha_1, \alpha_2, ...$*(finite or infinite sequence)* $\in A$ *with* $\alpha = \alpha_1 \odot \alpha_2 \odot ...\}$, thus, $A^{(\infty)} \cap \boldsymbol{PR}_0 =$
$\{\alpha \in \boldsymbol{PR}_0 \mid \exists \alpha_1, \alpha_2, ...$*(finite or infinite sequence)* $\in A$ *with* $\alpha = \alpha_1 \odot \alpha_2 \odot ...\} =$
$\{\alpha \mid \exists \alpha_1 \in A \cap \boldsymbol{PR}_0$ with $\alpha = \alpha_1\} = A \cap \boldsymbol{PR}_0$.

If L is ces-definable then $L \backsimeq PR\,[U]$ for a certain U and by (AA1) in Theorem 12.1:
$L \backsimeq \{pr(Q) \mid Q \in \boldsymbol{FC}_{\mathbb{X}}[U]\}^{(\infty)}$. Thus, $L \cap \boldsymbol{PR}_0 = \{pr(Q) \mid Q \in \boldsymbol{FC}_{\mathbb{X}}[U]\}^{(\infty)} \cap \boldsymbol{PR}_0 =$ (by (12.5)) $\{pr(Q) \mid Q \in \boldsymbol{FC}_{\mathbb{X}}[U]\} \cap \boldsymbol{PR}_0 = \{pr(Q) \mid Q \in \boldsymbol{FC}_{\mathbb{X}}[U]\}$.
Hence $(L \cap \boldsymbol{PR}_0)^{(\infty)} = \{pr(Q) \mid Q \in \boldsymbol{FC}_{\mathbb{X}}[U]\}^{(\infty)} \backsimeq L$.
Suppose that $L \backsimeq (L \cap \boldsymbol{PR}_0)^{(\infty)}$ and define
$U = \sum_{\alpha \in L} FOLD[\alpha]$. Let $\beta \in L$, $\beta \neq \mathbb{O}$, $\beta \neq \mathbb{I}$. Then $FOLD[\beta] \leq U$, hence $PR[FOLD[\beta]] \subseteq \boldsymbol{PR}[U]$. Since $\beta \in \boldsymbol{PR}[FOLD[\beta]]$, we get $\beta \in \boldsymbol{PR}[U]$. Thus, $L \subseteq \boldsymbol{PR}[U]$. Let $\beta \in \boldsymbol{PR}[U]$, $\beta \neq \mathbb{O}$, $\beta \neq \mathbb{I}$. By (AA1) in Theorem (12.1):
$\beta \in \{pr(Q) \mid Q \in \boldsymbol{FC}_{\mathbb{X}}[U]\}^{(\infty)}$. Hence, $\beta = \beta_1 \odot \beta_2 \odot ....$ - a finite or infinite concatenation
- where $\beta_j \in \{pr(Q) \mid Q \in \boldsymbol{FC}_{\mathbb{X}}[U]\}$ $(j = 1, 2, ...)$ and by (12.4):
$\beta_j \in \{pr(Q) \mid Q \in \bigcup_{\alpha \in L} \boldsymbol{FC}_{\mathbb{X}}[FOLD[\alpha]]\}$. This implies that there exists a process $\alpha_j \in L$ such that β_j is one of atomic processes which α_j factorises into. By assumption
$L \backsimeq (L \cap \boldsymbol{PR}_0)^{(\infty)}$ we get $\alpha_j \in (L \cap \boldsymbol{PR}_0)^{(\infty)}$, thus, $\alpha_j = \alpha_{j1} \odot \alpha_{j2} \odot ...$, for some

Fig. 12.1 Processes α and β

processes $\alpha_{j1}, \alpha_{j2}...,$ each belonging to $L \cap \boldsymbol{PR}_0$. Thus, all α_{jk} belong to L and are atomic processes. But among all these atomic processes, β_j appears (see Proposition 11.4(b)), thus $\beta_j \in L \cap \boldsymbol{PR}_0$. Since it holds for each $j = 1, 2, ...,$ we get $\beta = \beta_1 \odot \beta_2 \odot .... \in (L \cap \boldsymbol{PR}_0)^{(\infty)}$ and by assumption $L \backsimeq (L \cap \boldsymbol{PR}_0)^{(\infty)}$, we obtain $\beta \in L$. Therefore $\boldsymbol{PR}[U] \subseteq L$. Concluding, $L \backsimeq \boldsymbol{PR}[U]$, that is, L is ces-definable. □

The reader should notice that assumption (12.4) in Theorem 12.5 is essential indeed. Consider, for instance, the following process language containing two processes: $L = \{\alpha, \beta\}$ where α and β are depicted in Fig. 12.1.

Obviously $L \backsimeq (L \cap \boldsymbol{PR}_0)^{(\infty)}$, but (12.4) in Theorem 12.5 does not hold. Indeed, $FOLD[\alpha] = \{a_{x \bullet y}, b_y, x^a, y^{a \bullet b}\}$, $FOLD[\beta] = \{a_x, b_{x \bullet y}, x^{a \bullet b}, y^a\}$. Thus, $\boldsymbol{FC}_{\mathbb{X}}[FOLD[\alpha]] \cup \boldsymbol{FC}_{\mathbb{X}}[FOLD[\beta]] \neq \boldsymbol{FC}_{\mathbb{X}}[FOLD[\alpha] + FOLD[\beta]]$, since the set to the right of "$\neq$" contains, for instance, the firing component $\{a_x, x^a\}$, while the set to the left of "$\neq$"—does not. Thus, c-e structure $U = FOLD[\alpha] + FOLD[\beta]$ generates more processes than the language L contains. By (SS1) in Theorem 12.4, L is not ces-definable.

□

12.4 Some Classes of Process Languages

Consider the monoid of processes $\mathscr{M} = \langle \boldsymbol{PR}, \odot, \mathbb{I} \rangle$ (Corollary 11.1) and its power $\mathscr{S} = \langle \mathbb{P}(\boldsymbol{PR}), \cup, \odot, \{\mathbb{I}\} \rangle$, that is, a restriction of APL (Definition 12.1). Obviously, for any infinite chain of process languages $A_1 \subseteq A_2 \subseteq A_3 \subseteq ....$ and a process language B: $B \odot \bigcup_{i \geq 1} A_i = \bigcup_{i \geq 1} B \odot A_i$ (more general equation is (18)). In accordance with [3], any algebra similar to $\mathscr{S}$ with such infinite distribution property is called an *ω-complete semiring*. Let $\mathscr{X}$ be a (countable) set of variables, called *process language variables*. They will assume process languages as values. An example of such variable is in equations (14), (15). A *monomial* over $\mathscr{S}$ is an expression of the form $A_1 \odot A_2 \odot ... \odot A_m$ where $A_j \in \mathscr{X}$ or $A_j \subseteq \boldsymbol{PR}$ with $|A_j| < \infty$ $(j = 1, 2, ..., m)$. It does not restrict generality of considerations if $A_j = \{\alpha_j\}$ where $\alpha_j \in \boldsymbol{PR}$; such A_j is called a *constant*. For some variables $X_1, X_2, ..., X_n \in \mathscr{X}$, let $\overrightarrow{X} = (X_1, X_2, ..., X_n)$. A *polynomial* $p(\overrightarrow{X})$ over $\mathscr{S}$ is a finite union ($\cup$) of monomials. A system of *fix-point equations* is $X_i = p_i(\overrightarrow{X})$ for $i = 1, 2, ..., n$ where $p_i(\overrightarrow{X})$ are polynomials. A solution (or a fix-point of) to this system is an n-tuple of process languages $\overrightarrow{L} = (L_1, L_2, ..., L_n)$ such that $L_i = p_i(\overrightarrow{L})$. It is the least solution if for any n-tuple of process languages $\overrightarrow{L'} = (L'_1, L'_2, ..., L'_n)$ with $L'_i = p_i(\overrightarrow{L'})$, the inclusions $L_i \subseteq L'_i$ $(i = 1, 2, ..., n)$ hold. It follows from the general theory of semirings that the system $X_i = p_i(\overrightarrow{X})$ has a unique least solution and it is computed by the following iterative procedure:

$$\begin{cases} \overrightarrow{X}^{(0)} = (X_1^{(0)}, X_2^{(0)}, ..., X_n^{(0)}) = \underbrace{(\emptyset, \emptyset, ..., \emptyset)}_{n-times} = \overrightarrow{\emptyset} \\ \overrightarrow{X}^{(k+1)} = (p_1(\overrightarrow{X}^{(k)}), p_2(\overrightarrow{X}^{(k)}), ..., p_n(\overrightarrow{X}^{(k)})) \text{ for } k = 0, 1, 2, 3, ... \end{cases}$$

It may be checked that $X_i^{(k)} \subseteq X_i^{(k+1)}$ for $i = 1, 2, ..., n$ and each $k \geq 0$, which we denote $\overrightarrow{X}^{(k)} \leq \overrightarrow{X}^{(k+1)}$ ($\leq$ is a partial order). Thus, the least upper bound of the chain $\overrightarrow{X}^{(0)} \leq \overrightarrow{X}^{(1)} \leq \overrightarrow{X}^{(2)} \leq .....$ exists and it is the (least) solution to the system $X_i = p_i(\overrightarrow{X})$ [3].

By *algebraic* we call a general system of equations $X_i = p_i(\overrightarrow{X})$, $i = 1, 2, ..., n$, that is when $p_i(\overrightarrow{X})$ is an arbitrary polynomial of $\overrightarrow{X}$, as well as a process language differing at most by $\mathbb{O}$ or $\mathbb{I}$ from a solution to such system. The class of algebraic languages is denoted $ALG(\odot)$. Let $A, B \subseteq \boldsymbol{PR}$, $|A| < \infty$, $|B| < \infty$, $X \in \mathscr{X}$. By *linear* we call this system of equations if any monomial of $p_i(\overrightarrow{X})$ is of the form $A \odot X \odot B$ or A; also a process language differing at the most by $\mathbb{O}$ or $\mathbb{I}$ from a solution to such system is called linear. The class of linear languages is denoted $LIN(\odot)$. By *rational* we call this system of equations if any monomial of $p_i(\overrightarrow{X})$ is of the form $A \odot X$ or $X \odot A$ or A; also a process language differing at the most by $\mathbb{O}$ or $\mathbb{I}$ from a solution to such system is called rational. The class of rational languages is denoted $RAT(\odot)$. By Theorem 12.5, a process language L is ces-definable iff $L \backsimeq (L \cap \boldsymbol{PR}_0)^{(\infty)}$, provided that L satisfies the non-creativity Property 12.4. Denote by $CES(\odot)$ the class of ces-definable languages, each satisfying assumption (12.4) in Theorem 12.5 and containing finitely many atomic processes (Definition 10.5). Then we have:

Proposition 12.4 *$CES(\odot) \subset RAT(\odot) \subset LIN(\odot) \subset ALG(\odot) \subset \mathbb{P}(\boldsymbol{PR})$, with all the strict inclusions.*

Proof Only the first inclusion is worth a demonstration - the remaining are evident. Let $L \in CES(\odot)$. It suffices to find finite $A, B \subseteq \boldsymbol{PR}$ such that L solves equation $X = X \odot A \cup B$ or $X = A \odot X \cup B$. For $A = L \cap \boldsymbol{PR}_0$ (the finite set; $\boldsymbol{PR}_0$ is the set of all atomic processes) and $B = \{\mathbb{I}\}$ it is seen that $(L \cap \boldsymbol{PR}_0)^{\odot}$ solves these equations (see (14), (15)). Since $(L \cap \boldsymbol{PR}_0)^{(\infty)} = (L \cap \boldsymbol{PR}_0)^{\odot} \cup (L \cap \boldsymbol{PR}_0)^{(\omega)}$ (see (12.3)) and $(L \cap \boldsymbol{PR}_0) = (L \cap \boldsymbol{PR}_0) \odot (L \cap \boldsymbol{PR}_0)$, we conclude that $(L \cap \boldsymbol{PR}_0)^{(\infty)}$ solves these equations too. By Theorem 12.5 we get $L \backsimeq (L \cap \boldsymbol{PR}_0)^{(\infty)}$, hence L solves these equations up to $\{\mathbb{O}, \mathbb{I}\}$, thus $L \in RAT(\odot)$. Therefore $CES(\odot) \subseteq RAT(\odot)$. On the other hand, not every rational process language is ces-definable, since, for instance

$L = \{O, \ < a, 0 > \to < b, 1 > \to < c, 1 >, \ < b, 0 > \to < c, 1 >\} \in RAT(\odot)$ but $L \notin CES(\odot)$. Indeed, L solves the equation $X = \{< a, 0 > \to < b, 1 >\} \odot X \cup \{< b, 0 > \to < c, 1 >\}$, thus $L \in RAT(\odot)$ but $L \cap \boldsymbol{PR}_0 = \{< b, 0 > \to < c, 1 >\}$, thus $(L \cap \boldsymbol{PR}_0)^{(\infty)} = \{< b, 0 > \to < c, 1 >\}^{\odot} \cup \emptyset = \{\mathbb{O}, \mathbb{I}, < b, 0 > \to < c, 1 >\}$, thus $L \notin CES(\odot)$. □

Let us make two remarks.
1. Interpreting an equation $X_i = p_i(\overrightarrow{X})$ $(i = 1, 2, ..., n)$ as a collection of rewriting rules $X_i \to m_{ij}(\overrightarrow{X})$ where each $m_{ij}(\overrightarrow{X})$ $(j = 1, 2, ..., m)$ is a monomial of $p_i(\overrightarrow{X})$, one may treat the equation system as a grammar (with the process concatenation $\odot$ instead of ordinary word concatenation): context-free (for algebraic system), linear and regular (the latter for rational system) respectively. Then, a number of linguistic properties for such grammars may be derived.
2. The name "algebraic" (ω-language, equation system) is used following terminology in [1, 2] rather than e.g. in [6], where " algebraic" and "context-free" denote different classes of ω-languages.

References

1. Czaja L, Kudlek M (2000) Rational linear and algebraic process languages and iteration lemmata. Fundamenta Informaticae 43(1–4):49–60
2. Czaja L, Kudlek M (2001) ω-Process languages for place/transition nets. Fundamenta Informaticae 47(3, 4), 217–229
3. Golan JS (1992) The theory of Semirings with application in mathematics and theoretical computer science, Longman Scientific and Technical
4. Kleene CS (1956) Representation of events in nerve nets and finita automata, Automata Studies (eds. Shannon CE, McCarthy J) Princeton, pp 3–41
5. Reisig W (1985) Petri Nets. An introduction, number 4 in EATCS monographs on theoretical computer science, Springer, Berlin, Heidelberg, New York, Tokyo
6. Thomas W (1990) Automata on infinite objects, in handbook of theoretical computer science, volume B (Formal Models and Semantics), Jan van Leeuwen (editor), Elsevier

Chapter 13
Examples

The cause-effect structures may be used to specify, or model, various kinds of state transforming dynamic systems. These may be computer programs (algorithms expressed in a programming language), systems of real physical objects motion with rules of its control, systems where communication and synchronisation of actions plays a primary part, systems where coordination of actions must be ensured, their ordering desired or forbidden, systems of objects reacting on signals triggering off simultaneous action of a group of objects, etc. The following examples present samples of such issues.

13.1 Addition and Subtraction of Natural Numbers

Algorithms $[y := y + x;\ x := 0]$ and $[y := y - x;\ x := 0]$ with variables x, y of the non-negative integer type, are implemented by the c-e structures in Fig. 13.1. The variables are represented as nodes x and y, containing initially m and n of tokens respectively, will contain 0 and $n + m$ of tokens after termination of the addition, and $n - m$ if $n \geq m$—after termination of the subtraction. If $n \geq m$, the correct termination takes place (token at h) iff the variable y gets the final value and variable x gets the value 0. If $n < m$ then in case of subtraction, the termination takes place with a token at the node i and empty nodes a, h, y, whereas $m - n$ tokens at the node x. The node x plays both roles—inhibiting in the firing component $\{x^{\theta}_{\omega\otimes h}, i^{\theta}_{h}, h^{x\bullet i}_{\theta}\}$ and ordinary in the firing component $\{x^{\theta}_{a}, i^{\theta}_{a}, a^{x\bullet i}_{\theta}\}$ (*addition*) and $\{x^{\theta}_{a}, i^{\theta}_{a}, y^{\theta}_{a}, a^{x\bullet i\bullet y}_{\theta}\}$ (*subtraction*).

A decomposition into single arrows of the *addition* c-e structure in Fig. 13.1 is depicted in Fig. 13.2. Thus, this c-e structure in the "arrow-expression" representation (see Chap. 2) is $((i \to h) \bullet (x \to h)) + ((i \to a) \bullet (x \to a)) + (a \to y)) \bullet (a \to i)$.

L. Czaja, *Cause-Effect Structures*, Lecture Notes in Networks and Systems 45, https://doi.org/10.1007/978-3-030-20461-7_13

addition   *subtraction*

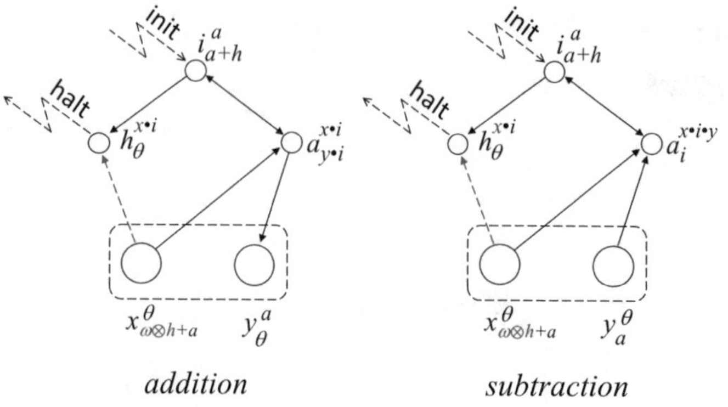

Fig. 13.1 If in the initial state s: $s(i) = 1$, $s(a) = 0$, $s(h) = 0$, and nodes x, y hold respectively $s(x) = m$, $s(y) = n$ tokens, then in the termination state t: $s(i) = 0$, $s(a) = 0$, $s(h) = 1$, and the nodes x, y hold $t(x) = 0$, $t(y) = n + m$ in case of *addition*, and $t(y) = n - m$ (provided that $n \geq m$) in case of *substraction*. The node x plays role of the inhibitor in the firing component $\{x_{\omega \otimes h}^{\theta}, i_h^{\theta}, h_{\theta}^{x \bullet i}\}$ and a role of the ordinary node in the firing component $\{x_a^{\theta}, i_a^{\theta}, a_{\theta}^{x \bullet i}\}$. Capacity of x and y is ω (infinite), capacity of i (initiation), h (halt), a (auxiliary) is 1. Thus, in accordance with Definition 3.4 (Chap. 3): $(s, t) \in [[addition]]^*$ and $(s, t) \in [[subtraction]]^*$

Fig. 13.2 Composition of the *addition* c-e structure out of single arrows

$$addition = \left(\begin{matrix} i_h^{\theta} & x_{\omega \otimes h}^{\theta} \\ \bigcirc & \bigcirc \\ \big\downarrow & \cdot & \big\downarrow \\ \bigcirc & \bigcirc \\ h_{\theta}^i & h_{\theta}^x \end{matrix} \right) + \left(\begin{matrix} i_a^{\theta} & x_a^{\theta} \\ \bigcirc & \bigcirc \\ \big\downarrow & \cdot & \big\downarrow \\ \bigcirc & \bigcirc \\ a_{\theta}^i & a_{\theta}^x \end{matrix} \right) + \left(\begin{matrix} a_y^{\theta} & a_i^{\theta} \\ \bigcirc & \bigcirc \\ \big\downarrow & \cdot & \big\downarrow \\ \bigcirc & \bigcirc \\ y_{\theta}^a & i_{\theta}^a \end{matrix} \right)$$

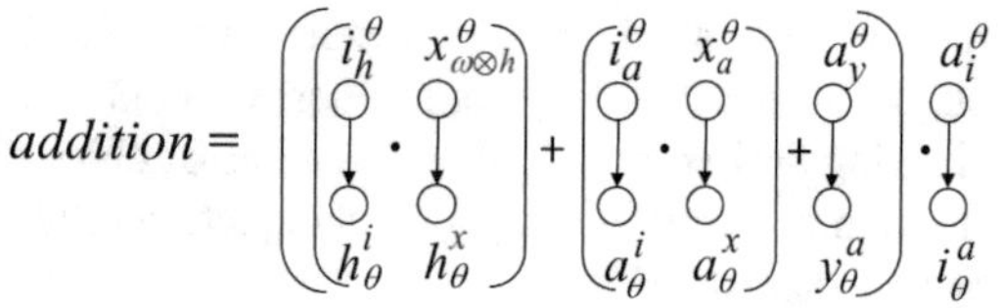

Remark C-e structures with some nodes of infinite capacity and with some nodes playing role of inhibitors, can implement computing with integers, each represented by a number of tokens. This is not so without inhibitors. This follows from the well-known fact in the Petri net theory and Chap. 9.

13.2 Producers and Consumers [1]

There are n producers $p[1]$, $p[2]$, ..., $p[n]$, m consumers $c[1]$, $c[2]$, ..., $c[m]$, and one wholesale house W. The work of a producer $p[i]$ consists in producing a commodity, delivering it to its dispatch agency $d[i]$, sending it to the wholesale house W and returning to production activity, and so on. The work of a consumer $c[j]$ consists in usage of the commodity, bring in next commodity from the wholesale house W to its reception agency $r[j]$ and returning to usage, and so on. The c-e structure specification of such Producers/Consumers system is depicted in Fig. 13.3.

The Producers/Consumers c-e structure in the "arrow-expression" representation is:

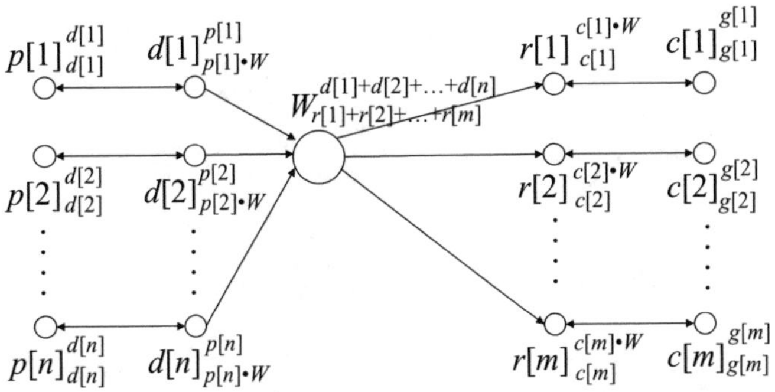

Fig. 13.3 Producers/Consumers. Capacity of the wholesale house W is ω (infinite) and of remaining nodes is 1. Initial state: empty nodes are $W, d[1], d[2], \ldots, d[n], r[1], r[2], \ldots, r[m]$ and token at $p[1], p[2], \ldots, p[n], c[1], c[2], \ldots, c[m]$

$$\sum_{i=1}^{n}\sum_{j=1}^{m} \overbrace{(p[i] \leftrightarrow d[i])}^{\text{producer}} \bullet \overbrace{(d[i] \rightarrow W \underbrace{\rightarrow r[j])}_{\text{recipient}}}^{\text{dispatcher}} \bullet \overbrace{(r[j] \leftrightarrow c[j])}^{\text{consumer}}$$

13.3 Cigarette Smokers' Problem [13]

There are three cigarette smokers, M, P, T and agent A, depicted in Fig. 13.4. The smokers have respectively: *M*atches, cigarette *P*apers and *T*obacco, but none of them can hand over his ingredient to another smoker. The Agent can supply any ingredient, so puts two of them at a time on the table in spots m (matches), p (paper), t (tobacco). A smoker lacking two ingredients picks them up from the table, makes a cigarette and smokes it. On completion, the smoker notifies the agent and waits for the next chance to smoke. The agent during a smoker's work is idle. Token at M, P, T means that respective smoker is smoking. Token at m, p, t means that respective ingredient is on the table. Token at A means that no one is smoking and there is no ingredient on the table. Unfortunately, a deadlock may occur: for instance when after the simultaneous moves $A \rightarrow m$ and $A \rightarrow p$, the move $p \rightarrow pM$ takes place.

Representation by "arrow expression" (see Chap. 2) of this c-e structure is:
$(A \rightarrow m \rightarrow mT \rightarrow T \rightarrow A) \bullet (A \rightarrow p \rightarrow pT \rightarrow T \rightarrow A)+$
$(A \rightarrow p \rightarrow pM \rightarrow M \rightarrow A) \bullet (A \rightarrow t \rightarrow tM \rightarrow M \rightarrow A)+$
$(A \rightarrow t \rightarrow tP \rightarrow P \rightarrow A) \bullet (A \rightarrow m \rightarrow mP \rightarrow P \rightarrow A)$
Prevention against deadlock without additional nodes is obtained by applying inhibitors, as demonstrates Fig. 13.5. The dashed arrows begin in the nodes playing both roles—ordinary and inhibiting (depending on firing components they belong to).

Remark Originally in [13], the cigarette smokers problem was an example of problems that cannot be programmed using ordinary (scalar) Dijkstra's [8] semaphores

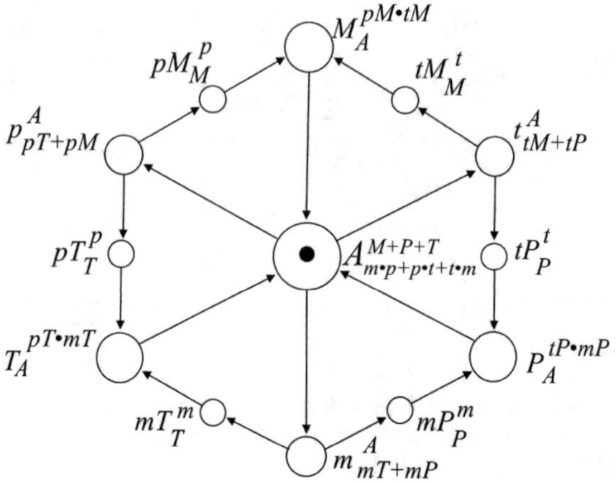

Fig. 13.4 Cigarette smokers deadlock-sensitive. Capacity of all nodes is 1. Initial state: token at A and empty remaining nodes

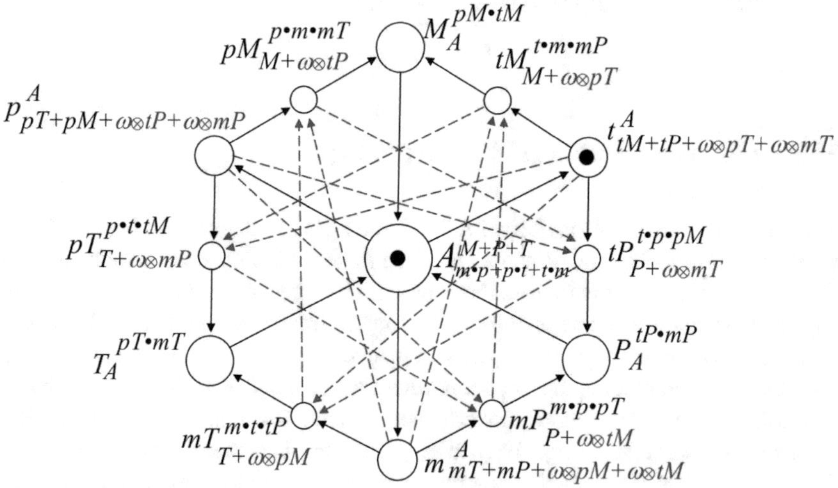

Fig. 13.5 Cigarette smokers deadlock-free. As in Fig. 13.4, the capacity of all nodes is 1 and initial state: token at A and empty remaining nodes

with no conditional statements as synchronisation means, so that a deadlock be avoided. A number of publications appeared thereafter, containing discussion of and solution to this problem, making proposals of stronger programming means, like vector semaphores, as well as variations and generalizations of the problem, some early ones, e.g. [10–12]. The phrase "cigarette smokers" is a metaphor of some synchronisation issues in operating systems.

13.4 Dining Philosophers [7]

Philosophers (originally 5) are sitting around a table, everyone sharing right and left fork with the right and left neighbour. Two forks are needed to eat fish, thus the philosophers must compete for the forks. Nodes $f[0], f[1], \ldots, f[n]$ $(n > 1)$ in the figures that follow represent forks: a token at $f[j]$ $(j = 0, 1, 2, \ldots, n)$ indicates presence of the jth fork on the table; nodes $e[0], e[1], \ldots, e[n]$ represent eating: a token at $e[j]$ indicates that the jth philosopher is eating; nodes $r[0], r[1], \ldots, r[n]$ represent uplift of the right forks and $l[0], l[1], \ldots, l[n]$—uplift of the left forks by respective philosophers: a token at $r[j]$ (resp. $l[j]$) indicates that the jth philosopher has just picked up the right (left) fork but is not using it yet. The story in the form of a decorated c-e structure is depicted in Fig. 13.6 for $n = 4$, i.e. for 5 philosophers. However, two deadlock states may occur: either when all philosophers pick up their left forks, or all pick up their right forks. Prevention from reaching the deadlocks is a task of a "butler" counting how many philosophers have uplifted their left forks and right forks. This is depicted in Fig. 13.7. The butler removes one token from the node L when a philosopher picked up left fork or one token from R when picked up right fork. The butler forbids a philosopher to pick up left fork when L is empty and to pick up right fork when R is empty. The butler returns tokens to L or R as soon as a respective philosopher starts eating. Initially, L and R contain n tokens.

Remark The early versions of this exercise appeared in [7–9]. In the solution presented here (and in several other published versions), the so-called "livelock" is not possible: it would occur if a philosopher uplifts a certain fork, puts it immediately back on the table, uplifts it again and so on endlessly. Such version, considered in some publications, results in the "starvation" of this philosopher's neighbours.

13.5 Crossroad

In Figs. 13.8, 13.9, 13.10 the c-e structures modelling traffic on a crossroad are depicted. Names of vertices are mnemonic: for instance, NS, EW, SN, WE are entry points to the crossroad from the north, east, south, west, resp., whereas ns, ew, sn, we—exit points to the south, west, north, east. So, $NS \to ew$ depicts the turn right from the north to the west, $NS \to W \to ns$—the straight ahead north-souths move, $NS \to N \to E \to we$—turn left from the north to the east, etc. Nodes N, E, S, W are points of collision. The collisions are prevented in Fig. 13.9 by means of inhibitors (nodes and arrows in red correspond to inhibition—prevention). Tokens at larger circles represent vehicles and at smaller circles—control signals indicating directions the vehicles arrive from. Let:

$X[1] = NS \to ew$	(north to west ride–right turn)
$Y[1] = NS \to W \to ns$	(north-south–straight ride)
$Z[1] = NS \to N \to E \to we$	(north to east ride–left turn)

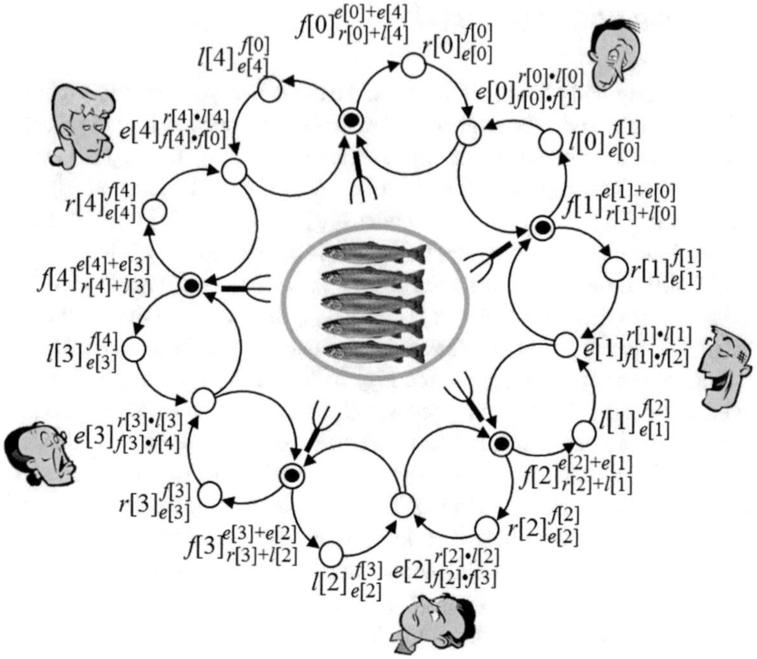

Fig. 13.6 Five dining philosophers with deadlock possible. The activity of the jth philosopher represented by an "arrow expression" is $(f[j] \to r[j] \to e[j] \to f[j]) \bullet (f[j \oplus 1] \to l[j] \to e[j] \to f[j \oplus 1])$ where $\oplus$ means addition modulo n, here $n = 4$, i.e. **if** $j < n$ **then** $j \oplus 1 = j + 1$ **else** $j \oplus 1 = 0$. Thus, the activity of their whole team is $\sum_{j=0}^{n} (f[j] \to r[j] \to e[j] \to f[j]) \bullet (f[j \oplus 1] \to l[j] \to e[j] \to f[j \oplus 1])$

$X[2] = EW \to sn$	(east-north ride–right turn)
$Y[2] = EW \to N \to ew$	(east-west–straight ride)
$Z[2] = EW \to E \to S \to ns$	(east- south ride–left turn)
$X[3] = SN \to we$	(south to west ride–right turn)
$Y[3] = SN \to E \to sn$	(south-north–straight ride)
$Z[3] = SN \to S \to W \to ew$	(south to east ride–left turn)
$X[4] = WE \to ns$	(west-south ride–right turn)
$Y[4] = WE \to S \to we$	(west-east–straight ride)
$Z[4] = WE \to W \to N \to sn$	(west-north ride–left turn)

Traffic from the north is modelled by $X[1] + Y[1] + Z[1]$, from the east—by $X[2] + Y[2] + Z[2]$, from the south—by $X[3] + Y[3] + Z[3]$ and from the west—by $X[4] + Y[4] + Z[4]$—respective denotations are in Fig. 13.8. The arrow expression specifying physical traffic, however enabling some unlawful ride is: $\sum_{i=1}^{4} (X[i] + Y[i] + Z[i])$.

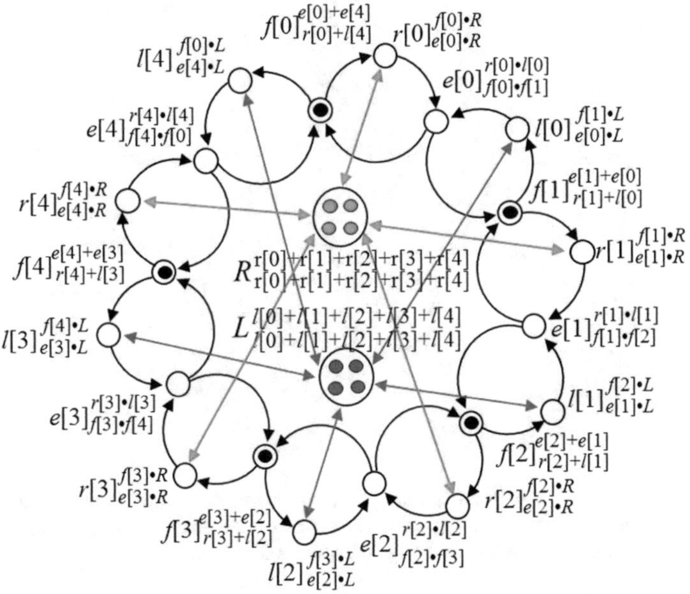

Fig. 13.7 Five dining philosophers with deadlock prevented by the butler managing left forks (L) and right forks (R)

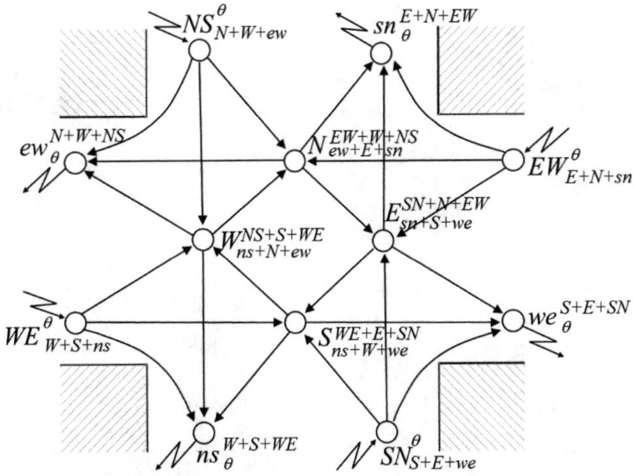

Fig. 13.8 Crossroad—physical traffic with some unlawful moves, for instance $NS \to N \to sn$, $NS \to N \to ew$, $NS \to W \to N$, $NS \to W \to ew$, etc

More traffic regulations is imposed by introducing alternating ride permit in mutually orthogonal directions, in respective Figures—north-south and east-west. To this end, an oscillator with min-time (Chap. 4), $V \leftrightarrow H$ is combined with the c-e structure in Fig. 13.9, as shown in Fig. 13.10. A token in V (red light for vertical traffic)

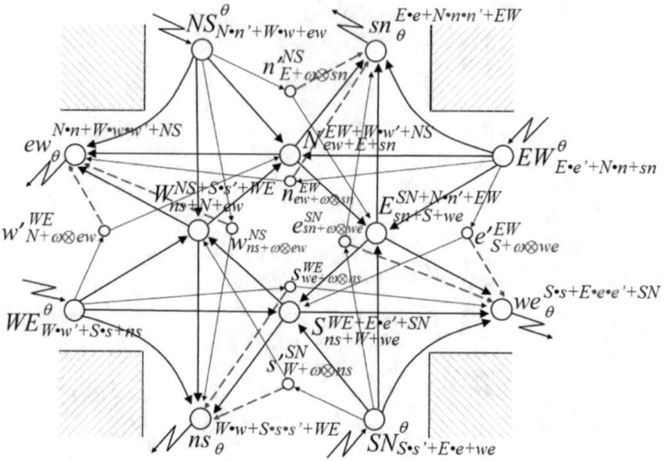

Fig. 13.9 Crossroad—physical traffic with all lawful moves. Prevention against unlawful moves is obtained by applying inhibitors. The dashed red arrows begin in the nodes playing both roles—ordinary and inhibiting (depending on firing components they belong to)

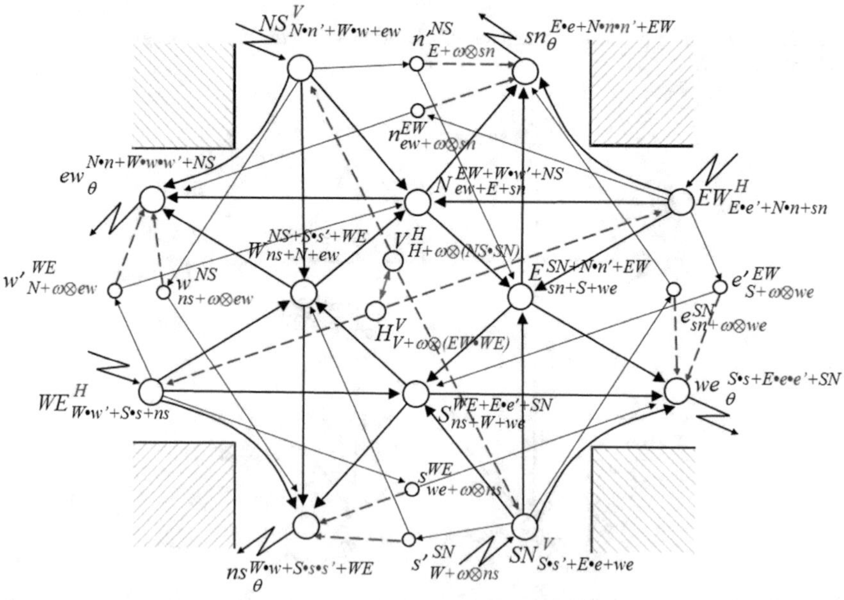

Fig. 13.10 Crossroad—physical traffic with lawful motion additionally controlled by the min-time oscillator $V \leftrightarrow H$, where a token at V represents red light for the vertical traffic while in H—for the horizontal

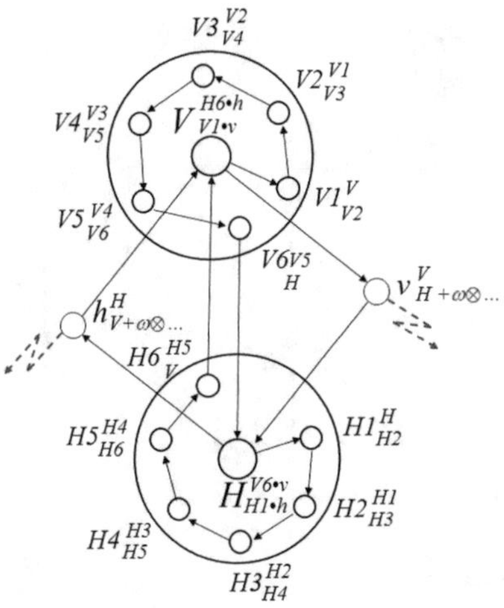

Fig. 13.11 A no-time c-e structure simulating the min-time c-e structure $V \leftrightarrow H$ with $T_{min}(V) = T_{min}(H) = 6$. The dashed zigzag red arrows go out of nodes, which play both roles—ordinary and inhibiting. These arrows lead to nodes where vertical and horizontal traffic respectively begins on the crossroad. The initial state: a token (i.e. red light) either in V or in H but not in both

inhibits entry vehicles into NS and SN, whereas token in H (red light for horizontal traffic) inhibits entry vehicles into EW and WE. The oscillator, a simple min-time c-e structure, may be simulated by means of no-time c-e structure as shown in Fig. 13.11, where $T_{min}(V) = T_{min}(H) = 6$ time units.

Remark As a measure of the crossroad throughput, a number of vehicles passing it in a certain time period, can be admitted. This depends on the traffic intensity from both directions, but also on the frequency of the red light oscillation between V and H. The crossroad throughput may be controlled by modification of the $T_{min}(V)$ and $T_{min}(H)$ values, for a given traffic intensity. To the min-time oscillator $V \leftrightarrow H$ may be straightforwardly added the intermediate "yellow lights", as shows Figs. 13.12 and 13.13.

13.6 Road Grid

Interconnecting the cross-roads from Fig. 13.8 and further, a grid of roads may be obtained, an example of which is depicted in Fig. 13.14. To avoid excessive entanglement of arrows, some nodes internal to the cross-roads are omitted. The visible nodes are indexed by their position in rows and columns in the grid. The subscripts and superscripts of node names are omitted. The grid can be extended arbitrarily in both directions.

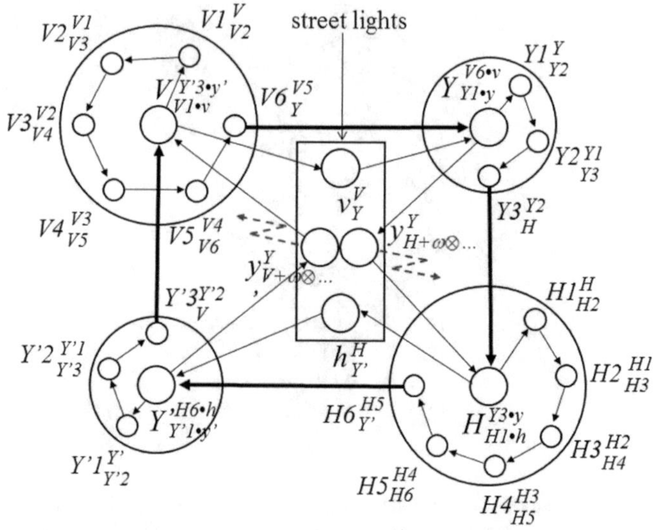

Fig. 13.12 A no-time c-e structure simulating the min-time c-e structure $V \to Y \to G \to Y' \to V$ with $T_{min}(V) = T_{min}(H) = 6$, $T_{min}(Y) = T_{min}(Y') = 3$, which models red, yellow and green street lights, where a token at V means red light for the vertical traffic, while in H—for the horizontal traffic

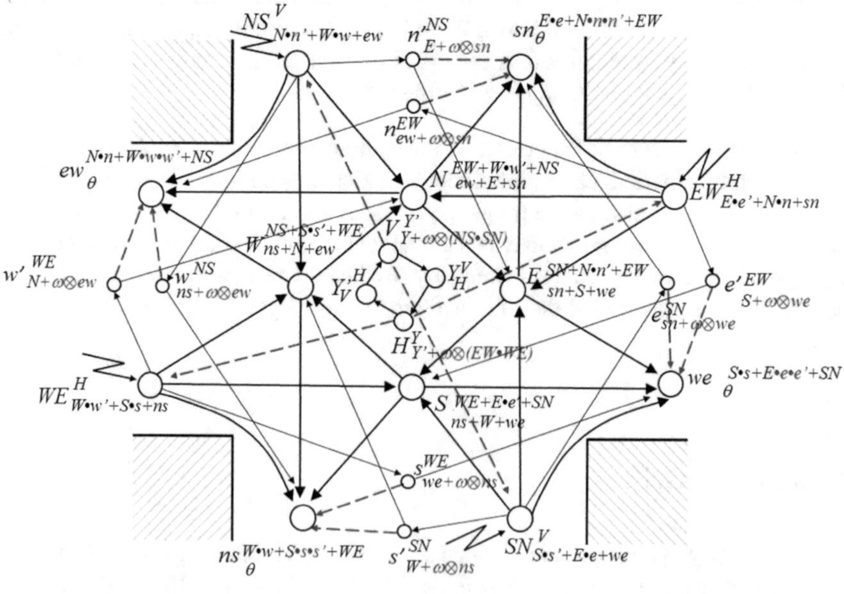

Fig. 13.13 Crossroad—physical traffic with lawful motion additionally controlled by the min-time c-e structure $V \to Y \to G \to Y' \to V$, which models red, yellow and green light—dependendly on a token presence at one of these nodes

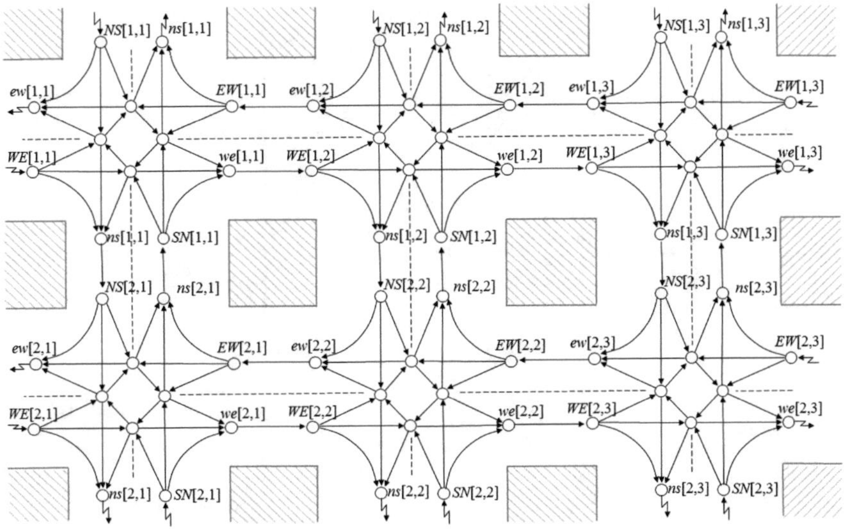

Fig. 13.14 Road grid—a "city". The traffic proceeds along the arrows, observing rules specified in Fig. 13.13

13.7 Lifts

There are two lifts (elevators), left and right, in a 3-storey house, depicted as c-e structure in Fig. 13.15 in the "bus-layout" (see the drawinng patterns in Fig. 2.4, Chap. 2), where nodes $1A$, $2A$, $3A$ represent stop-points of the left lift and $1B$, $2B$, $3B$—stop-points of the right lift, in respective lift shafts. The system has no control mechanism for the ordinary usage, it shows only physical motion of the lifts without any restrictions. It is assumed that a token, representing a lift, is held by exclusively one node $1A, 2A, 3A$ at a time and that the same concerns the nodes $1B, 2B, 3B$.
In the "arrow expressions" (Chap. 2) the system may be specified as:

$$LEFTLIFT = (1A \leftrightarrow 2A) + (2A \leftrightarrow 3A) + (3A \leftrightarrow 1A)$$
$$RIGHTLIFT = (1B \leftrightarrow 2B) + (2B \leftrightarrow 3B) + (3B \leftrightarrow 1B)$$
$$LIFTS = LEFTLIFT + RIGHTLIFT$$

A control mechanism imposed on the physical motion comprises a pair of additional nodes $\langle j, j' \rangle$ (where $j, j' = 1, 2, 3$) on the staircase of each storey, and largely uses inhibitors, which prohibit some moves. The Fig. 13.16 depicts a system (perhaps somewhat old-fashioned) of two lifts with this the control mechanism. The nodes j and j' represent push-buttons: calling a lift and getting into it, respectively. Inserting a token into the node j means a lift-call to the jth-storey, that is pressing the call-button represented by the node j. This is inhibited, thus impossible, if both lifts are on the jth storey. If at least one lift is on this floor, then a token may be inserted into node j', meaning that a user gets in to a lift and may chose a storey k as a destination. This results in move of this token to a node kA or kB, for $k \neq j$ with simultaneous move

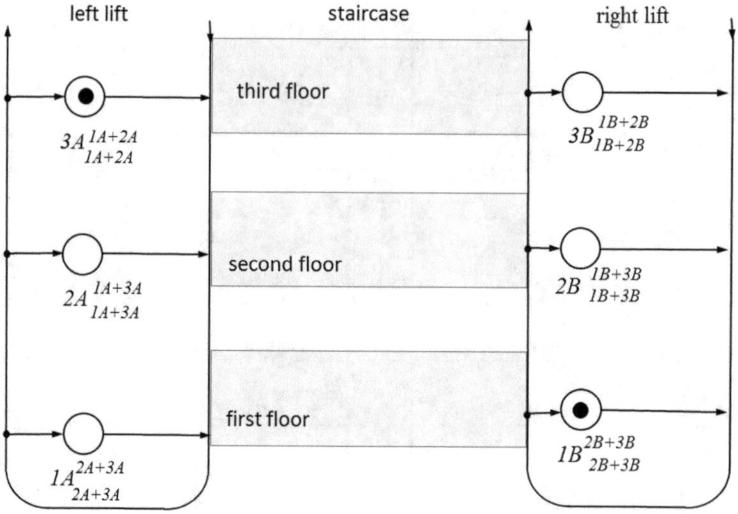

Fig. 13.15 C-e structure in the "bus layout", depicting physical motion of lifts, with the left lift on the third floor and the right—on the first

of token (a lift) from jA to kA or from jB to kB. Evidently, a token, representing a lift, can be at exactly one stop-point at a time, meaning presence of this lift on a respective storey. The dashed red arrows make inhibit of some token moves.

Remark Various modifications of the system model depicted in Fig. 13.16 are possible. For instance, call a lift to a storey if another lift is on this storey (possible in the presented model) might be prevented by applying additional inhibitors. The model may be straightforwardly extended to arbitrary number of lifts and storeys.

13.8 Alternating Bit Protocol [3]

This is a "packet switching" solution to the Alternating Bit Protocol (ABP) introduced by K.A.Barlett, R.A.Scantlebury and P.T.Wilkinson in 1969. The solution presented here is a slight modification of the c-e structure that models the ABP, published in [5]. The task is to transmit messages and acknowledgments between sender and recipient through *unreliable* channels, so that: (a) messages and acknowledgments reach destinations in the order of their sending, (b) messages and acknowledgments lost in the channels are duplicated and resent, (c) every message and acknowledgment reaches its destination exactly once. A message apart from the first, is not dispatched before acknowledgement of reception of the preceding message arrives. A simplified (unrealistic) model, which assumes reliable channels is depicted in Fig. 13.17, the complete (realistic) model, with unreliable channels—in Fig. 13.18. The models use variables B and A assuming bit values ($1 = \textbf{true},\ 0 = \textbf{false}$), that accompany,

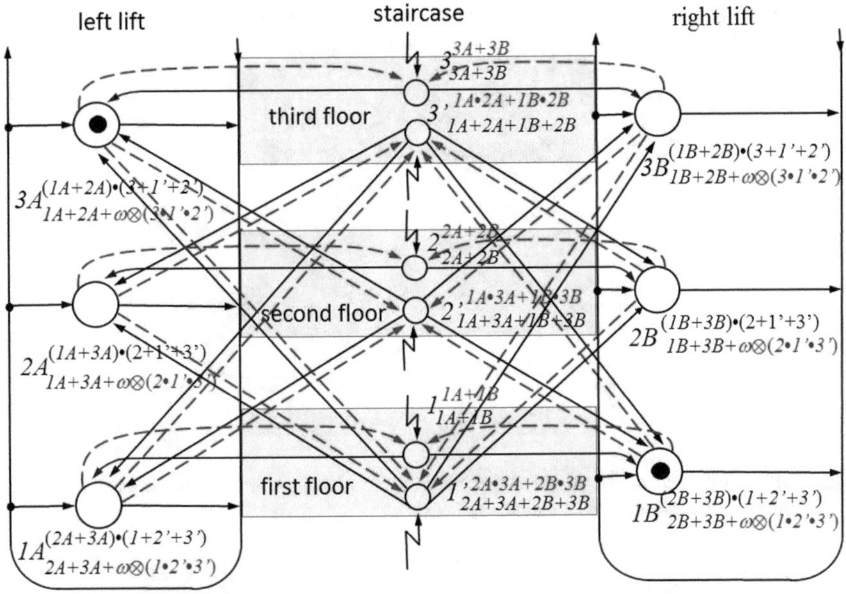

Fig. 13.16 Two lifts system with the left lift (a token at node *3A*) on the third floor, and right lift—on the first (a token at node *1B*). Node $j = 1, 2, 3$ represents a lift call button in the staircase. Insertion of a token into node j, means pressing the button, whereas moving this token to the position jA or jB means arrival of a lift into the jth storey. If both lifts are already on the jth storey, then insertion of this token is not possible (due to nodes jA and jB playing role of inhibitors in this case), that is, pressing button j brings no effect. If at least one lift is on the jth storey a token may be inserted into the node j'. This means entering the lift, impossible when no lift is on the storey j (precluded by respective inhibitors). Dislocation of the token to the lift's stop-point kA or kB (for $k \neq j$) means movement of the lift to storey k

respectively, message MES and acknowledgment ACK in packets travelled round-trip between the sender and recipient. Some arrows rooted in nodes are annotated by comments (green) that refer to actions performed in these nodes, in particular they show values of bits B and A assumed when traversing these arrows by tokens. A packet $\langle MES, B\rangle$ travels through the channel from m to q, then arrives in node u or r (on the recipient's side), depending on value of B and state of the "flip-flop" $t \leftrightarrow s$, unless lost or deleted during the travel. The deletion takes place if B's value differs from the value determined by the flip-flop's current state: 1 if t holds a token and 0 if s does. The same concerns a packet $\langle ACK, A\rangle$ travelling from z to d, then to e or h (on the sender's side), where the current state of flip-flop $f \leftrightarrow g$ determines 1 if h holds a token and 0 if e does. The flip-flops' state corresponds to a phase of circulation of data throughout the protocol from message dispatch to acknowledgment reception. In every phase of the circulation the recipient and sender should get identical value of A and B, alternately 1, 0, 1, 0, 1, 0, etc.- in successive correct transmissions.

In the "arrow expressions" (Chap. 2), the ABP model shown in Fig. 13.18 may be specified as follows:

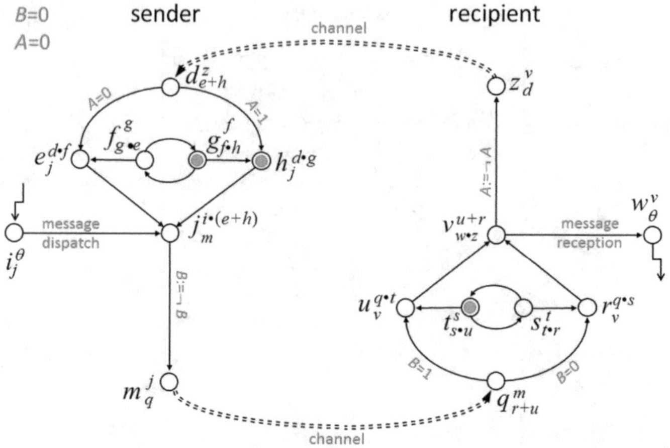

Fig. 13.17 Simplified model of the ABP: reliable (faultless) channels thus no duplication and retransmission of messages and acknowledgments

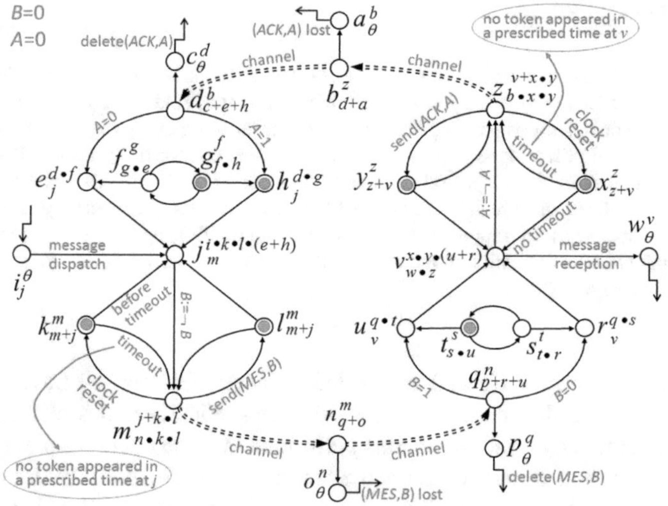

Fig. 13.18 Complete model of the ABP: unreliable (faulty) channels, thus duplication and retransmission of messages and acknowledgments necessary

$MESCHAN = (m \to n \to q) + (n \to o) + (q \to p)$ (channel for message)
$ACKCHAN = (z \to b \to d) + (b \to a) + (d \to c)$ (channel for acknowledgment)
$TAKEMES = (i \to j) \bullet (k \to j) \bullet (l \to j) \bullet [(e \to j) + (h \to j)]$
$SENDMES = (m \to l) \bullet (m \to k) \bullet (l \to m) \bullet (k \to m)$
$MESALT = j \to m$
$ACKFLIPFLOP = (f \to e) \bullet (f \to g) \bullet (g \to f) \bullet (g \to h)$
$ACKSKIP = d \to c$
$ACKLOST = b \to a$
$RECEIVEACK = (b \to d) \bullet [(d \to e) + (d \to h)] \bullet ACKFLIPFLOP + ACKSKIP + ACKLOST$

$TRANSMIT = SENDMES \bullet MESCHAN + MESALT + RECEIVEACK \bullet TAKEMES$
$DELIVERMES = (v \to w) \bullet (x \to v) \bullet (y \to v) \bullet [(r \to v) + (u \to v)]$
$SENDACK = (z \to y) \bullet (z \to x) \bullet (y \to z) \bullet (x \to z)$
$ACKALT = v \to z$
$MESFLIPFLOP = (t \to u) \bullet (t \to s) \bullet (s \to t) \bullet (s \to r)$
$MESSKIP = q \to p$
$MESLOST = n \to o$
$RECEIVEMES = (n \to q) \bullet [(q \to r) + (q \to u)] \bullet MESFLIPFLOP + MESSKIP + MESLOST$
$RECEIVE = SENDACK \bullet ACKCHAN + ACKALT + RECEIVEMES \bullet DELIVERMES$
$ALTERNATINGBITPROTOCOL = TRANSMIT + RECEIVE$

Remark The actions on annotated arrows (commented in green in Figs. 13.17, 13.18), can be implemented by c-e structures located in nodes where these arrows are rooted. For instance, time expiry (timeout) of dispatch of message or acknowledgment, may become realized by a maxtime c-e structure (see 4.2, Chap. 4), whereas testing of value of bit variables B and A may be accomplished by suitably applied c-e structure named "test zero", depicted in Fig. 3.5, Chap. 3. The ABP protocol is a design pattern for its various versions, modifications and extensions. Some referred to as "sliding window" [4] or "Abracadabra" [2], have been applied, for instance in the known TCP protocols.

13.9 Volley—Enforcement of Simultaneous Actions

A troop of shooters represented by nodes $0, 1, \ldots, n$ ($n = 4$ in Fig. 13.19) should simultaneously fire a shot. Any shooter (perhaps with two firearms) can notify his/her two neighbours by poking them only—no vocal and visible command was issued for all, and no time of the shot was specified. The initiator (node 2 in the Figure) starts the poking process, which propagates throughout all the troop. The auxiliary node $n + 1$ (5 in the Figure) rebounds the wave of signals (pokes) propagating from the initiator. The volley may be fired only when all the shooters get signals, i.e. tokens. This is represented by simultaneous move of these tokens from $0, 1, \ldots, n$ to $1', 2', \ldots, [n + 1]'$. Notice that this local communication between the shooters is enforced by using at most two operands in monomials (products) of the upper and lower polynomials associated with each node's name.
This c-e structure may be specified by the following "arrow expression":

$$\sum_{j=0}^{n+1}(j \to [j \ominus 1]) \bullet (j \to [j \oplus 1]) + \bigwedge_{j=1}^{n}([j \to j') \bullet (j \to [j \oplus 1]') \bullet (0 \to 1') \bullet (0 \to [n+1]')$$

where $j \oplus 1 = \begin{cases} j+1 & \text{if } j < n+1 \\ 0 & \text{if } j = n+1 \end{cases}$ $\quad j \ominus 1 = \begin{cases} j-1 & \text{if } j > 0 \\ n+1 & \text{if } j = 0 \end{cases}$

and where $n = 4$ for the c-e structure depicted in Figure 13.19.

Remark If the wave of pokes starts from the shooter different than 2, then the system in Fig. 13.19 would run into a deadlock—it would never come to the simultaneous

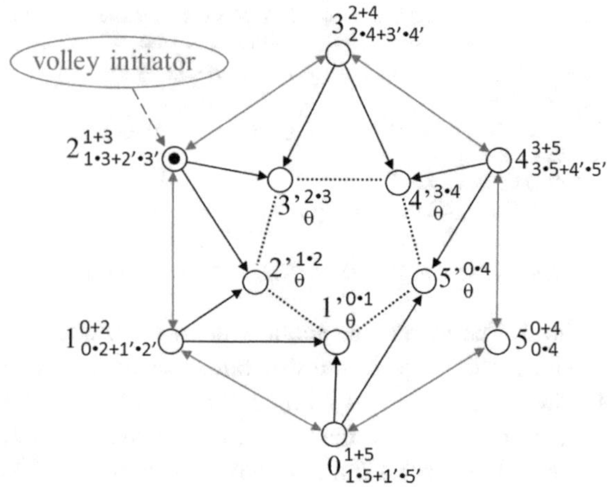

Fig. 13.19 A c-e structure modeling the volley fired at the same instant by shooters 0, 1, 2, 3, 4. Node 5 rebounds the signals' wave, started at node 2. Bidirectional brown arrows represent poking of neighbours by the shooters

shot. Notice that in the arrow expression, the repetitive sum $\sum$ of c-e structures specifies propagation of signals (pokings) along the shooters, whereas the repetitive product $\bigwedge$—their simultaneous shot.

13.10 Boatman–A Coordination Problem

This early medieval puzzle has been analysed in terms of Petri nets in [6]. The puzzle touches a problem of coordination of actions, the enforcement of their permissible or forbidden order in particular. Here it is specified and solved in terms of c-e structures with inhibitors, as show Figs. 13.20 and 13.21. A boatman (b) should carry across the river a goat (g), wolf (w) and cabbage (c) having room for one of them (beside himself) in his boat. To avoid dangerous events, that is eating up the cabbage by the goat or the goat by the wolf when the boatman is away, he must cross the river several times from one bank to the other, taking care not to leave alone wolf and goat or goat and cabbage on the same river bank when he is not there. A token at c, b, g, w represents presence of respective participant of the river crossing, on one river bank, and at c', b', g', w'—on the other. There are dangerous initial moves in Fig. 13.20: from b to b', from b along with c to b' and c', as well as from b along with w to b' and w', and a number of others while crossing the river in both directions. All these prohibited moves are eliminated by imposition of inhibitors as shows Fig. 13.21.

This insecure c-e structure is specified by the following "arrow expression":

$(b \to b') \bullet [(b \to g') + (b \to c') + (b \to w')]\bullet$

$(g \to g') \bullet (c \to c') \bullet (w \to w') \bullet (b' \to b) \bullet (g' \to b) \bullet (g' \to g)$

Secure transport of all passengers to the opposite river bank is specified by the c-e structure in Fig. 13.21.

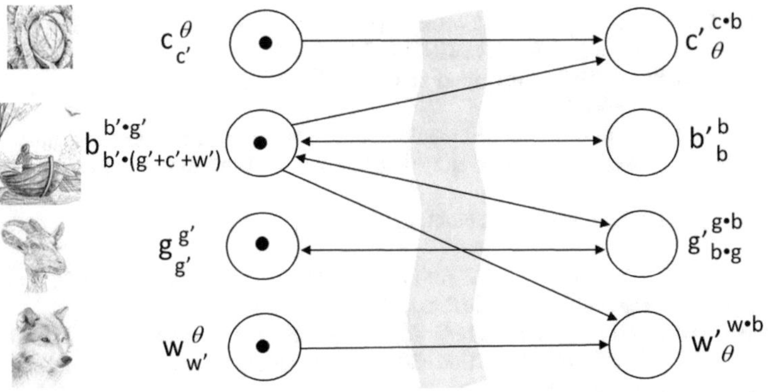

Fig. 13.20 Physical movements of the boatman and his passengers without restrictions; the cabbage and goat endangered of being eaten

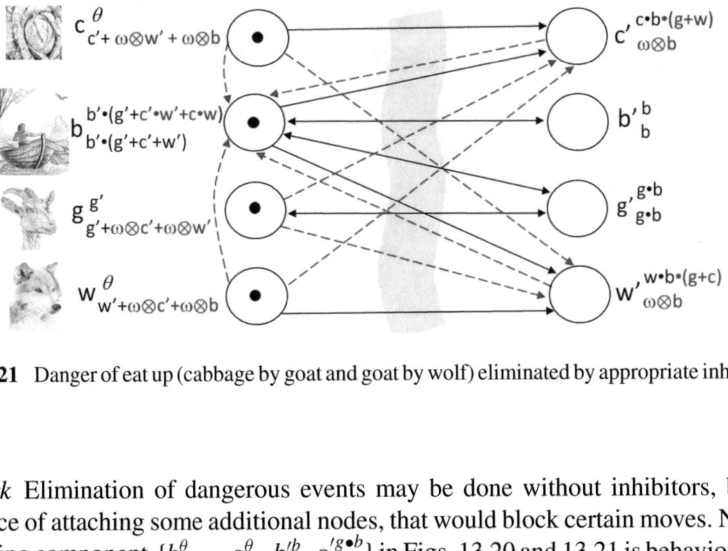

Fig. 13.21 Danger of eat up (cabbage by goat and goat by wolf) eliminated by appropriate inhibitors

Remark Elimination of dangerous events may be done without inhibitors, but at the price of attaching some additional nodes, that would block certain moves. Notice that firing component $\{b^{\theta}_{b'\bullet g'}, g^{\theta}_{g'}, b'^{b}_{\theta}, g'^{g\bullet b}_{\theta}\}$ in Figs. 13.20 and 13.21 is behaviourally equivalent to closely connected firing component $\{b^{\theta}_{b'\bullet g'}, g^{\theta}_{g'\bullet b'}, b'^{b\bullet g}_{\theta}, g'^{g\bullet b}_{\theta}\}$ (remind the Remark after Definition 2.8 in Chap. 2), which contains more arrows. The same concerns the firing component $\{b'^{\theta}_{b}, g'^{\theta}_{g\bullet b}, b^{b'\bullet g'}_{\theta}, g^{g'}_{\theta}\}$, equivalent to $\{b'^{\theta}_{b\bullet g}, g'^{\theta}_{g\bullet b}, b^{b'\bullet g'}_{\theta}, g^{g'\bullet b'}_{\theta}\}$. Usage of firing components not closely connected, reduces redundant arrows.

References

1. Brinch HP (1973) Operating system principles. Prentice-Hall, Inc., Englewood Cliffs, New Jersey, USA
2. Broy M, Střlen K (2001) Abracadabra protocol, specification and development of interactive systems, Monograps in Computer Science

3. Barlett KA, Scantlebury RA, Wilkinson PT (1969) A note on reliable Full-duplex Transmission over Half-duplex links. Commun ACM 12(5):260–261
4. Comer, Douglas E (1995) Internetworking with TCP/IP, Volume 1: Principles, Protocols, and Architecture, Prentice Hall
5. Czaja L (2018) Introduction to distributed computer systems. Principles and features. Springer
6. Desel J (1985) Another Boatman story, Petri nets and related system models, Newsletter 22, October 1985
7. Dijkstra EW (1965) Solution of a problem in concurrent programming control. Commun ACM 8/9
8. Dijkstra EW (1968) In: Genuys F (ed) Co-operating sequential processes, in programming languages, Genuys edn. Academic Press, New York
9. Dijkstra EW (1971) Hierarchical ordering of sequential processes. Acta Inf 1:115–138
10. Habermann AN (1972) On a solution and a generalization of the cigarette smokers' problem, Technical Report, Carnegie-Mellon Univ
11. Parnas DL (1972) On a solution to the cigarette smokers' problem (without conditional ststements), Internal report
12. Parnas DL (1975) On a solution to the cigarette smokers' problem (without conditional ststements), Commun ACM 18
13. Patil SS (1971) Limitations and capabilities of Dijkstra's semaphore primitives for coordination among processes, Project MAC, Computational Structures Group Memo 57

Bibliography

Related Literature not Referenced in the Chapters

On Cause-Effect Structures

1. Bialasik M (1992) On a logic for cause-effect structures. MSc thesis (in Polish), Warsaw University
2. Bialasik M (1994) Cause-effect structure logic. Institute of Computer Science Polish Academy of Sciences, Internal Report
3. Ciobanu G (1998) Conditional seminear-rings as cause-effect structures. Cuza University, Romania, Technical Report
4. Czaja L (1988) Specification by cause-effect structures. In: Proceedings of the IFIP W.G.2.3 24th meeting, Zaborow, 19–23 June 1988
5. Czaja L (1989) Finite processes in cause-effect structures and their composition. Inf Process Lett 31
6. Czaja L (1990) Modelling systems by cause-effect structures. In: Invited paper to the 3rd seminar on modelling, evaluation and optimization of dependable computer systems, Berlin, Wendisch-Rietz, Nov 1990 (in 2nd vol of Proceedings)
7. Czaja L (1992) Some semantics of cause-effect structures and their relationships. In: Proceedings of the workshop concurrency, specification and programming, Berlin, Nov 1992
8. Czaja L (1993) Net-models for abstract data types. In: Proceedings of the workshop concurrency, specification and programming, Nieborów, Oct 1993
9. Czaja L (1995) Lattice of cause-effect structures and their set-theoretic representation. Institute of Informatics, Warsaw University TR 95-06(206)
10. Czaja L (1995) Behaviour of cause-effect structures (Properties of four semantics). Institute of Informatics, Warsaw University TR 95-10(210)
11. Czaja L (1995) Representing CSP-like systems as cause-effect structures (revised). In: Proceedings of concurrency, specification and programming workshop, Warsaw, Oct 1995, pp 55–76
12. Czaja L (1996) Cause-effect structure processes and their link with Mazurkiewicz traces. Institute of Informatics, Warsaw University TR 96-03(224)
13. Czaja L (1996) Minimal/maximal time cause-effect structures (revised version). Institute of Informatics, Warsaw University TR 96-07(228)
14. Czaja L (1996) Processes in cause-effect structures. In: Bjorner D, Broy M, Pottosin IV (eds) Perspectives of system informatics, second international Andrei Ershov memorial conference, Akademgorodok, Novosibirsk, Russia, June 1996. Lecture notes in computer science, vol 1181
15. Czaja L (1997) Examples of specification by cause-effect structures (30 case studies). Institute of Informatics, Warsaw University TR 97-03(240)

L. Czaja, *Cause-Effect Structures*, Lecture Notes in Networks and Systems 45, https://doi.org/10.1007/978-3-030-20461-7

16. Czaja L (1998) Cause-effect structures-structural and semantic properties revisited. Fundamenta Informaticae 33:17–42
17. Czaja L (1998) Minimal/maximal time cause-effect structures (revised). Fundamenta Informaticae 33:1–16
18. Czaja L (1999) Representing hand-shake channel communication in the calculus of cause-effect structures. Fundamenta Informaticae 37(4):343–368
19. Czaja L (2000) Elementarnye priczinno-sledstwiennye struktury. Sistemnaya informatika 7, Nauka 2000, pp 7–81 (preliminary and partial version of the present text, translated into Russian)
20. Czaja L, Maggiolo-Schettini A (1992) Notes on time cause-effecte structures. Dipartimento di Informatica Universita di Pisa, Internal notes
21. Dung Tran Viet (1994) Liveness in cause-effect structures. MSc thesis (in Polish), Warsaw University
22. Dusza M (1992) Creation of concurrent programs controlled by cause-effect structures. MSc thesis (in Polish), Warsaw University
23. Grygiel K, Weiss Z (1993) Generalized markovian cause-effect structures. In: Proceedings of the workshop concurrency, specification and programming, Nieborów, Oct 1993
24. Holenderski L, Szalas A (1988) Propositional description of finite cause-effect structures. Inf Process Lett 27:111–117
25. Komarzynska E (1995) A graphic interface to a language and interpreter of concurrent programs based on cause-effect structures. MSc thesis (in Polish), Warsaw University
26. Lisa S (1993) Equivalenze per structure causa-effetto. Tesi di laurea in Scienze dell'Informazione, Universita di Pisa
27. Matteucci G (1993) Processi in Structure Causa-Effetto. Tesi di laurea in Scienze dell'Informazione, Universita di Pisa
28. Maggiolo-Schettini A, Matteucci G (1993) Processes of cause/effect systems. Technical Report: TR-34/93, Universita degli Studi di Pisa Dipartimento di Informatica
29. Maggiolo-Schettini A, Matteucci G (1997) Processes in cause/effect systems. Fundamenta Informaticae 31:305-335
30. Nguyen Duc Ha (1991) Reference manual of the CAD for cause-effect nets (in Polish). RP.I.09 Project, Internal Report, Institute of Informatics, Warsaw University
31. Nguyen Duc Ha (1991) Cause-effect structures and Peti nets. PhD thesis (in Polish), Institute of Informatics, Warsaw University
32. Puchalski M (2001) Multivalued cause-effect structures. MSc thesis (in Polish), Warsaw University
33. Raczunas M (2000) Algebra procesów generowanych przez struktury przyczyn i skutków (Algebra of processes generated by cause-effect structures). Submitted as a PhD thesis (in Polish), Warsaw University
34. Tran DV (1994) Liveness in cause-effect structures. MSc thesis (in Polish), Warsaw University
35. Ustimenko AP (1993) Modelling of cause-effect structures of Czaja in terms of regular petri nets. Problemy Teoreticheskogo I Eksperimentalnogo Programirowania, Nowosibirsk (in Russian)
36. Ustimenko AP (1993) Algebra of generalized cause-effect structures. Institute of Informatics Russian Academy of Science Report, Nowosibirsk
37. Ustimenko AP (1994) Mapping of time cause-effect structures into time regular Petri nets, in specification. In: Nepomniaschy VA (ed) Verification and net models of concurrent systems, Novosibirsk
38. Ustimenko AP (1996) Algebra of two-level cause-effect structures. Inf Process Lett 59:325–330
39. Ustimenko AP (1997) Cause-effect structures and Petri nets: relationship and comparative analysis. PhD thesis (in Russian), Russian Academy of Sciences, Sibirian Branch, Institute of System Informatics
40. Ustimenko AP (1998) Coloured cause-effect structures. Inf Process Lett 68(5):219–225
41. Weiss Z, Grygiel K (1992) Stochastic cause-effect structures: a simple model. In: Proceedings of the workshop concurrency, specification and programming, Berlin, Nov 1992

42. Wiech W (1995) Cause-effect structures with finite velocity of signal propagation (in Polish). MSc thesis, Warsaw University
43. Wikarski D (1993) Further remarks of the relationships between c-e structures and Petri nets. In: Proceedings of the workshop concurrency, specification and programming, Nieborów, Oct 1993

Other Literature

44. Baeten J, Weijland WP (1990) Process algebras, vol 18. Cambridge Tracts in Theoretical Computer Science
45. Barlett KA, Scantlebury RA, Wilkinson PT (1969) A note on reliable full-duplex transmission over half-duplex links. Commun ACM 12(5):260–261
46. Berthomieu B, Diaz M (1991) Modelling and verification of time-dependent systems using time Petri nets. IEEE Trans Softw Eng 17(3)
47. Berthomieu B, Menasche M (1983) An enumerative approach for analysing time Petri nets. In: Mason REA (ed) Proceedings IFIP 1983, North-Holland, pp 41–467
48. Best E, Fernandez C (1988) Non-sequential processes. A Petri net view, number 13 in EATCS monographs on theoretica computer science. Springer, Heidelberg
49. Best E, Devillers R, Hall JG (1992) The box calculus: a new causal algebra with multi-label communication. In: Advances in Petri nets. Lecture notes in computer science, vol 609, pp 21–69
50. Busi N, Gorrieri R (1995) A Petri net semantics for π*-calculus*. In: Proceedings of CONCUR 1995. Lecture notes in computer science, vol 962, pp 145–159
51. Coolahan JE, Roussopoulos N (1983) Timing requirements for time-driven systems. IEEE Trans Softw Eng SE-9(5)
52. Czaja L (1997) Process languages and nets. In: Proceedings of the workshop concurrency, specification and programming, Warsaw. (Also, to appear in Theoretical Computer Science)
53. Czaja L (2000) Process languages and nets. Theor Comput Sci 238:161–181
54. Czaja L (2003) On the analysis of Petri nets and their synthesis from process languages. RAIRO Theor Inf Nad Appl 37:17–38
55. Czaja L, Kudlek M (2000) Rational, linear and algebraic process languages and iteration lemmata. Fundamenta Informaticae 43(1–4):49–60
56. Czaja L, Kudlek M (2000) ω-process languages for place/transition nets (extended abstract). In: Proceedings of concurrency, specification and programming, vol 1, Berlin, 9–11 Oct 2000
57. Czaja L, Kudlek M (2001) Processes in place/transition nets, process languages and iteration lemmata. Submitted to RAIRO
58. Dijkstra EW (1976) A discipline of programming. Prentice-Hall Inc., Englewood Cliffs, New Jersey, USA
59. Degano P, De Nicola R, Montanari U (1988) A distribured operational semantics for CCS based on condition/event systems. Acta Informatica 26:59–91
60. Golan JS (1992) The theory of semirings with application in mathematics and theoretical computer science. Longman Scientific and Technical
61. Goltz U (1988) On representing CCS programs as finite Petri nets. In: Proceedings of MFCS 1988. Lecture notes in computer science, vol 324, pp 339–350
62. Goltz U, Mycroft A (1984) On the relationship of CCS and Petri nets. In: ICAPL 1984. Lecture notes in computer science, vol 172, pp 196–208
63. Goltz U, Reisig W (1984) CSP-programs as nets with individual tokens. In: Advances in Petri nets. Lecture notes in computer science, vol 188, pp 169–196
64. Gruska DP, Maggiolo-Schettini A (1993) Timed network semantics for communicating processes. In: Proceedings ot the concurrency, specification and programming workshop, Nieborow, Poland, Oct 1993, pp 125–143

65. Hansen ND, Madsen KH (1983) Formal semantics by a combination of denotational semantics and high-level Petri nets. In: Pagoni A, Rozenberg G (eds) Applications and theory of Petri nets, Informatik-Fachberichte, vol 66. Springer, Heidelberg, pp 132–148
66. Hoare CAR (ed) (1990) Developments in concurrency and communication. Addison-Wesley Publishing Co., Inc
67. Inmos Ltd. (1984) OCCAM programming language. Prentice Hall
68. Jensen K, Schmidt EM (1985) PASCAL semantics by a combination of denotational semantics and high-level Petri nets. In: Advances in Petri nets. Lecture notes in computer science, vol 222, pp 297–329
69. Kotov VE (1978) An algebra for parallelism based on Petri nets. In: LNCS 64. Springer, Heidelberg, pp 39–55
70. Kotov WE (1984) Seti Petri, Moskva, "Nauka" Glawnaya Redakcya Fiziko-Matematitcheskoy Literarury (in Russian)
71. Kuich W, Salomaa A (1986) Semirings, automata, languages. EATS monographs on theoretical computer science, vol 5. Springer, Berlin
72. Mazurkiewicz A (1987) Trace theory. In: Brauer W et al (eds) Petri nets. In: Applications and relationship to other models of concurrency, no 255. Lecture notes in computer science. Springer, Heidelberg, pp 279–324
73. May D, Shepherd R (1988) The transputer inplementation of OCCAM. Prentice Hall, UK
74. Messeguer J, Montanari U, Sassone V (1992) On the semantics of place/transition Petri nets. Universita Degli Studi di Pisa, Dipartimento di Informatica, Technical Report: TR-27/92
75. Nivergelt J, Farrar JC, Reingold EM (1974) Computer approach to mathematial problems. Prentic-Hall Inc., Englewood Cliffs, New Jersey, USA
76. Olderog E-R (1987) Operational Petri net semantics for CCSP. In: Advances in Petri nets. Lecture notes in computer science, vol 266, pp 196–223
77. Olderog E-R (1989) Nets, terms and formulas: three views of concurrent processes and their relationship. Institut fuer Informatik und Praktishe Mathematik Christian-Albrechts-Universitat zu Kiel, Habilitation dissertation
78. Petri CA (1977) Non-sequential processes. Internal Report GMD-ISF-77-5, Gesellschaft fuer Mathematik und Datenverarbeitung, Bonn
79. Petri CA (1982) State-transition structures in physics and in computation. Int J Theor Phys 21(10/11)
80. Popova-Zeugmann L (1991) On timed Petri nets. J Inform Process Cybern EIK 27(4):227–244
81. Popova-Zeugmann L (1992) Petri nets with time restricted places. In: Proceedings of the concurrency, specification and programming workshop, Berlin, pp 100–112
82. Schneider FB (1991) A seminar talk given at computer science department, Ben-Gurion University, Israel
83. Starke P (1980) Petri netze, Grundlagen, Anwendungen, Theorie. VEB Deutscher Verlag der Wissenschaften, Berlin
84. Starke P (1990) ATNA-arc timed net analyser. In: Petri Net Newsletter 37
85. Winkowski J (1982) An algebraic description of system behaviours. Theor Comput Sci 21:315–340
86. Winskel G (1985) Categories of models for concurrency. In: Seminar on concurrency. Lecture notes in computer science, vol 197, pp 246–267
87. Winskel G (1987) Petri nets, algebras, morphisms and computationality. Inf Comput 72:197–238
88. Zuberek WM (1980) Timed Petri nets and preliminary performance evaluation. In: Proceedings of the 7th annual symposium on computer architecture, May 1980, La Baule, France, pp 88–96

Index

L. Czaja, *Cause-Effect Structures*, Lecture Notes in Networks and Systems 45, https://doi.org/10.1007/978-3-030-20461-7

Printed in the United States
By Bookmasters